Human Agency in a Digital World

Human Agency in a Digital World

*/ Understand technology and
make it work for you*

< MARCUS FONTOURA

8080
BOOKS

an imprint of Microsoft

Distributed and printed by Simon & Schuster

Published in the United States by 8080 Books,
an imprint of Microsoft Corporation

1 Microsoft Way, Redmond, WA 98052

https://aka.ms/8080books

Cover Design by Shyla Lindsey
Typesetting by Nord Compo
In association with Wise Wolf Creative

ISBN 979-8-9997550-2-5 (Paperback – 8080 Books)
ISBN 979-8-9997550-3-2 (Hardcover – 8080 Books)
ISBN 979-8-9997550-4-9 (eBook – 8080 Books)

MICROSOFT, 8080 BOOKS, and the 8080 BOOKS Logo
are trademarks of Microsoft Corporation

To my family, especially my daughters, Ayesha and Shanti, whose curiosity about technology and productivity inspired me to write this book.

Endorsements

"As someone responsible for building some of the largest cloud systems on the planet, I've seen firsthand how technology can feel like it's racing ahead of us all. *Human Agency in a Digital World* is the antidote: Marcus Fontoura doesn't just explain AI, quantum and cloud with clarity and wit—he hands you the keys and shows you how to drive. This book is a practical, empowering guide for anyone who wants to use technology as a tool (not a magic trick), and a timely reminder that it's our human agency—our choices, curiosity, and values—that truly shape the future. In an age where AI headlines can make even the experts pause, Marcus gives everyone—from engineers to executives—the confidence and practical know-how to steer, not just spectate. Read it, enjoy it, and get ready to take the wheel."

Girish Bablani, President at Microsoft Azure

"With the rising importance of software and artificial intelligence in our world, it's increasingly important for everybody to understand how this new technology works, to demystify algorithms, and to democratize understanding of computer systems. Marcus Fontoura's *Human Agency in a Digital World* does a fantastic job of explaining complex concepts in simple terms, to make computer science topics approachable for anybody who would otherwise be intimidated to learn."

Hadi Partovi, Founder and CEO at Code.org

"Marcus wrote a witty and creative book about how information systems and AI works. The use of examples and tales that people can easily relate to will help readers quickly build intuition in a broad sense, from what is possible and real with AI to its energy efficiency! This kind of intuition gives people agency in a world where information systems and AI affect every aspect of our lives. This book is a delight to read, and can only be written by someone like Marcus who actually deeply understands the space, makes factual claims, and doesn't get people lost in noise and hyperbole going around about AI."

Prof. Luis Ceze, Lazowska Endowed Professor,
Paul G. Allen School of Computer Science
& Engineering, University of Washington

"As someone who has built real-world systems, from large-scale search engines to AI-driven healthcare platforms, I found *Human Agency in a Digital World* to be both timely and refreshingly grounded. The book argues that understanding technology is no longer a niche technical skill but a form of modern literacy, essential for navigating everything from algorithms to social infrastructure.

What distinguishes this book is its lens: it frames computing not as magic, but as a vehicle for efficiency and human agency. It sidesteps both techno-optimism and fear-mongering, focusing instead on how digital systems, search, cloud, AI, even quantum computing, can augment human capabilities, but only if guided by intentional, values-driven choices. The taxonomy of physical systems, physically-backed myths, and digital-only myths is especially insightful, and is a mental model I now find myself using when evaluating both technologies and organizations.

The writing is accessible, often witty, and never talks down. From the cautionary tale of NASA's Mars Climate Orbiter to the absurdity of bogosort, the book distills technical depth into vivid, memorable stories. If anything, the breadth occasionally comes at the expense of depth in later chapters, but that tradeoff is purposeful. This isn't a technical manual, it's a call to understand enough to steer.

For technologists, educators, policymakers, and curious citizens alike, *Human Agency in a Digital World* is not just a guide to the systems shaping our lives, it's a reminder that those systems remain ours to shape."

Kira Radinsky, Cofounder and CTO
at Diagnostic Robotics

"Imagine a book where you learn how complex things work through stories, movie characters, books, and memories drawn from your own imaginary world. That book exists. *Human Agency in a Digital World,* by Marcus Fontoura, invites you on a journey where imagination meets insight, making the digital age feel personal, accessible, and alive.

Tracing the evolution of digital technologies—from the earliest concepts of computing to the rise of artificial intelligence—this book helps readers understand that technology isn't magic, but the product of human ingenuity. Behind every digital artifact are scientists, engineers, and creators whose work drives progress and shapes the world we live in.

At a time when digital technologies and algorithms deeply influence our lives, Marcus Fontoura's book offers timely insight into their dilemmas, complexities, and promises. I commend the author for his inclusive and engaging approach to unveiling the intricacies of technology for the lay reader. This book sits at the heart of today's most pressing debates about the digital world."

Prof. Virgilio Almeida, Faculty Associate at Berkman Klein Center for Internet and Society, Harvard University. Author of *Algorithmic Institutionalism*

"Human Agency in a Digital World is an authoritative and engaging exploration of our digital world—from algorithmic efficiency and digital transformation to the interplay of physical systems and digital myths. Marcus Fontoura masterfully bridges deep technical insight with remarkable clarity, even making the quantum computing sections accessible and compelling. It stands as a significant contribution to the public's understanding of these complex systems."

Prof. Matthias Troyer, Technical Fellow at Microsoft

"Information technology is having both a positive and negative impact on everyone on this planet. This book provides a fascinating overview of this effect by making knowledge of computer technology and its impact on society accessible to a broad segment of society."

Prof. Donald Cowan, Distinguished Professor Emeritus of Computer Science, University of Waterloo, Canada

"What Marcus Fontoura has done with *Human Agency in a Digital World* is an intellectual tour-de-force illuminating both the origins of our digital transformation, the patterns behind these changes, and the way our lives might evolve to embrace what might initially be a more complicated future, but ultimately becomes a simpler form of abundance. The way Marcus draws analogies between diverse topics such as Daniel

Kahneman's work, computer programming, and neural networks exposes something rare: it shows us how the greatest systems thinkers who have been powering the transformation to the internet, mobile, the cloud, search, and now AI, have been actively fusing concepts from everywhere.

In the same way large language models have slurped up all the information on the web, deep thinking distinguished engineers at the top technology companies have used the power of the web to explore concepts far and broad, fusing them back into products and technologies to benefit our lives. No where is that on better display than in this book. You will find yourself coming back over and over again to sections and chapters as the web of intellectual connections unveils itself."

Brad Porter, Founder and CEO of Cobot.
Formerly Distinguished Engineer for Amazon
leading robotics and CTO for Scale AI.

Table of contents

Foreword

There are many good books written by visionaries, business leaders and academics that bravely and sometimes brilliantly speculate about techno utopias and anti-utopias. My favorite by far is Mustafa Suleyman's *The Coming Wave*. Compared to those books, *Human Agency in a Digital World* stands out in several ways. Its author, Marcus Fontoura, is an engineer to the core of his being. At his day job Marcus works on the most technically complicated layer of Microsoft's infrastructure. Prior to that, he was building web search infrastructure at Google. His team of over a thousand engineers creates technologies at the foundation of Microsoft's public cloud, Azure, which powers not only Microsoft services, ranging from Outlook to Bing search, but also most Fortune 500 companies.

Some executives are engineers and some engineers are executives. Marcus belongs to the second category. He is one of the most prominent technical experts at Microsoft. Being an engineer is not just a job, it is a mindset and a culture. Marcus explains how complex algorithms and systems work, sometimes adding insightful and humorous pop culture references. A lot of writing about technology is sugar syrup brewed by influencers—engaging and easy to digest but only marginally nutritious. This book is different. It is written by an engineer with a capital E, and it shows. It is a more complex read than almost anything one could find on a bookshelf of an airport newsstand. However, per unit of mental effort, reading *Human Agency in a Digital World* has exceptionally high ROI (return on investment) in terms of developing the reader's mind and understanding of the world of technology. It is an excellent read for anybody considering careers in engineering or business. Anybody who is intellectually curious and not afraid to strain their mind would benefit from this book. Someone with an algebra phobia can skip more technical parts and still get a lot out of the book and someone with graduate training in computer science might skip the "obvious" parts and benefit from the rest. There is only one prerequisite for reading this book: willingness to stretch the mind and desire to do so. Although the book is intended for a general audience, it is a great reading for a college level course about technology and society. It is particularly suitable for business thought leaders because it contains a wealth of examples from the world of business.

We tend to marvel at newly invented technologies, and we are keenly aware of their potential to transform society. Yet, we take for granted technologies that have been around for decades, noticing them only in the rare moments when they fail to work as expected. It is easy to forget how technologies from the very recent past transformed not just the economy, but our lives. This book does not dwell on the future – it is grounded at the present. Most people neither understand nor appreciate the complexity of multilayered technical infrastructure required for making a simple phone call or saving a file "in the cloud," and very few of us have keen awareness how different we would be without these technologies. The author does a great job helping the reader appreciate the technological marvels that we tend to take for granted providing accessible yet substantive explanations about how algorithms and technologies ranging from search and AI to public cloud actually work.

For instance, most of us never ask ourselves why video streaming and email fail so infrequently. Medieval cathedrals do not fail even after centuries of neglect because the walls are massive. One might naively, and erroneously, assume that email is reliable for somewhat similar reasons. Internet services running on components that frequently fail – yet, the service continues to work smoothly because the system is cleverly engineered to be robust to failure – there is an interesting analogy here not with an old building that are reliable because the walls are so strong but with organizations: when some employees get sick or resign unexpectedly, a well-functioning organization continues to deliver the service or product without interruptions.

Only in the digital age engineers learned to build systems that have the kind of robustness that life forms and well-functioning organizations exhibit. Why do new digital technologies tend to be unreliable, gaining reliability over time, and what are the tricks that engineers discovered to improve the reliability of complex systems over time? One will find the answer to this and many other fascinating questions in this book.

Michael Schwarz, Microsoft Chief Economist
August 2025

Introduction

While putting the finishing touches on my previous book, *A Platform Mindset*, I began to take a step back and think less about the power of digital platforms, and more about the power that we hold as individual users of these technologies, as people. This human power, which exists at both a personal and organizational level, is apparent whether looking at AI use today or quantum computing capabilities in the future.

Increasingly, the content we consume is focused on what technology is doing for us or to us, as if we are just passive bystanders. But each and everyone of us, regardless of our technical background or understanding of technology, has **human agency**. What does that mean? While there are many definitions, each makes it clear that we humans are autonomous beings, meaning that we act independently. As human agents, don't we owe it to ourselves to understand the strengths and weaknesses of AI and technology we use in our daily lives? My answer is "yes, absolutely."

I am a computer scientist, but I am a human first. A human who has daughters, to whom I've dedicated this book, and who wants to build a better world for them and for future generations. I care deeply about both technology and people, and as a result, how they intersect and impact one another. That's the first big idea of this book: human agency, which is our greatest tool to shape the impact of technology in our lives. Humans should understand technology. Evaluate it. Use it, if it is empowering, and not use it, if it is harmful. This concept is based on the definition of machine usefulness (MU) from MIT economists and Nobel laureates Daron Acemoglu and Simon Johnson's insightful book *Power and Progress*.[1] More important than being intelligent, efficient, or anything else, computer systems should above all enhance the human experience.

The second big theme of this book is something we often take for granted: how to use gains in **efficiency** to land a positive impact in society. The term "efficiency" today often rings like a 1950s corporate term for things running smoothly and economically. That is true, in part, but this term is woefully in need of modernization for today's day and age, especially as it is used in the context of today's technologies. Computers were first developed to improve the efficiency of numerical computations, and eventually evolved to create efficiencies in virtually all aspects of human life. Essentially, over time, we have been able to do more with less. It is my contention that the problems we face as a society, such as global climate change, hunger, healthcare, growth and productivity, at their core, are problems of efficiency. These complex

---

1. *Power and Progress: Our Thousand-Year Struggle Over Technology and Prosperity,* D. Acemoglu and S. Johnson, PublicAffairs, 2023.

problems yearn for efficiency-oriented solutions such as more efficient manufacturing, agriculture, pharmacology, and industries.

This book is the meditation of a computer scientist. I've set out to write a self-help book for this and future digital generations. How do we develop our human agency even as we become increasingly more technologically efficient? I'm a big believer that human agency and technological efficiency must coevolve if we want to create a better world. Everyone should understand intelligent technologies and develop their unique human abilities and capacities. This is more important than ever before. I hope this book will provide some of the foundations needed to have these discussions, and that it will inspire a new generation of leaders and technologists to understand the intersectionality of technology and society.

* * *

"If you can't explain something to a first-year student, then you really haven't understood."

Richard Feynman

It's no news to anyone that our lives are becoming ever more dependent on technology. We are in the middle of the AI revolution and on the verge of the quantum computing revolution. The better we understand the technologies shaping our lives today, the better positioned we will be to guide future technology use. This book was born out of my intent to explain the key concepts behind computer systems, with the hope of demystifying our fear of algorithms and helping us understand how they can be instrumental in creating a brighter future, with a more equitable and prosperous society.

It may be difficult for many of us to realize how much of our daily activities are simplified, amplified, or sometimes degraded, by computer systems. When we listen to music while driving to work, how many systems are we interacting with? If somehow the system is glitchy, and we cannot play our Spotify playlists, what is really happening? I wanted to give people the tools to understand what is going on under the covers. A Caltech faculty once asked physicist and Nobel laureate, Richard Feynman, to explain a difficult concept in theoretical physics. After thinking about the problem for a few days, he came up with the opening quote for this book.[2] Of course he also said that if you think you understand quantum mechanics, you don't understand quantum mechanics! Given our current dependency on computer systems, and the expectation that such dependency will only increase, I want to explain the foundational elements to enable people to understand technologies such as social media, web search, cloud computing, AI, and quantum computing.

Computer technology was primarily born to enable efficiency. The Manhattan project employed human

2. https://magazine.caltech.edu/post/feynman-at-100 (Last accessed in March 2025).

computers, which were ordinary women and men, to do calculations. Before the first computers were operational, this was common practice. The Electronic Numerical Integrator and Computer (ENIAC), which was the first programmable digital computer, was ten thousand times faster than the human computers employed by the Manhattan Project. The efficiency in computation brought by the ENIAC played a crucial role in the Cold War, where it was used for various projects, including development of the hydrogen bomb. Since then, the continued development of computer technologies has widened the gap between human and digital computers. The efficiencies produced by computer systems have allowed us to achieve more in faster and more accessible ways. We have better and faster news access. We shop online. We have cybersecurity systems. The current AI revolution, for instance, has only been possible due to the abundance of computational resources enabled by cloud computing and due to the large amounts of data generated or captured by computer systems. As I will argue in this book, computers are merely capable of making computations more efficient. Computers alone are not able to increase efficiency in society. We need human creativity and agency to transform computational efficiencies into real-world positive impacts on humankind.

Important questions about the future of humanity hinge on efficiency tradeoffs. Are we going to push for a society that focuses on doing more with less above all else, or will we consider how technology can be used to augment our humanity? South Korea exemplifies a high-efficiency society shaped by post-war urgency, Confucian values, and a tech-driven economy. The country boasts world-class infrastructure, ultra-fast internet, and a highly disciplined education and work culture that fuel innovation and global cultural influence. However, this relentless focus on speed and productivity comes at significant social costs, including long work hours, extreme academic pressure, mental health challenges, and the world's lowest birth rate. While South Korea's coordinated systems and national cohesion have enabled rapid progress, the country now faces a critical balancing act between maintaining its global edge and ensuring societal well-being and sustainability.[3]

We can automate boring, tedious tasks we don't like doing. Computers are great at those, and they also don't complain about doing the same tasks over and over. Computers can also make us smarter. The newer generations of chess players are likely better than those that couldn't play against computers, or any opponent for that matter, anytime they wished. Other uses of technology, unfortunately, can negatively impact society. The effect of social networks on teenagers' mental health may be the most evident example. The initial intent and excitement of being able to efficiently connect people across the globe ignored important tradeoffs. Of course, it may

---

3. https://www.ft.com/content/6e70f7bd-e311-41df-94fe-7a5575493ae6 (Last accessed in May 2025).

not have been possible to anticipate the consequences of social networks before they were used, but it can be used as a lesson learned going forward. While I don't claim to resolve every issue here, I hope to offer a useful point of view, including historical perspectives. The more we understand about the underpinnings of technologies that power the computer systems that impact our lives, the more agency we gain to shape them in ways that serve us. This is the basis of human agency.

I have spent most of my career working on the infrastructure for search engines and cloud computing services. Many of the projects I worked on are related to efficiency, answering questions like: how to make search queries faster? How to design search engines to use fewer computational resources? How to democratize cloud computing use by lowering their costs? How to make cloud computing more sustainable by using energy more efficiently? And how to design an organization to execute these projects in a cost effective way? These efficiencies are enablers for us to tackle the world's most relevant problems. By managing computational resources better, cloud providers enabled fundamental advances in AI, which in turn allowed us to make progress on basic science at a faster rate. As an anecdotal example, the 2024 Nobel Prize in chemistry was awarded to researchers working on AI to advance our understanding of proteins and predicting their structures. Such capabilities can be utilized for research and development that are fundamental for understanding biological mechanisms and the treatment of complex diseases, such as Alzheimer's disease.

I tried to present the concepts and algorithms described in this book in simple terms, with the intent to reach a wider audience, including readers that don't have a computer science background but are interested in learning more about the topic. I hope this book will inspire people to try to understand computer systems at a deeper level. Since I could not cover every possible computer system in a single book, I chose to focus on a few technologies that have, or will likely have, a huge impact on our lives: social media, web search, cloud computing, AI, and quantum computing. I strived to present the technical concepts in an increasing order of difficulty. Starting with simple algorithms, our journey goes through the avenues of sorting, compression, distributed systems, and NP-hard problems. Chapter 1 starts by defining some basic concepts which are used throughout the book. After that, each chapter introduces a new technology, building on the concepts previously described.

A foundational way to understand technology begins with recognizing that all computer systems are ultimately about transforming data into useful output through efficient computation. First, we must understand the nature and structure of the data available, including its sources, limitations, and representations. Given that data is often massive and noisy, we normally have to use techniques like compression and abstraction to reduce complexity and highlight signal over noise. This prepares the data for algorithms, which are structured procedures, or recipes, for extracting meaning, solving problems, or generating responses to difficult questions. The efficiency of these algorithms matters deeply, as inefficient systems consume more energy and resources without improving outcomes. Once algorithms are selected, they are executed within systems, like cloud platforms, social networks, search engines, which must be engineered for scalability and robustness.

So, the theory goes like this. If there is a problem to solve:

- Understand the data and process it, for instance compressing it and creating appropriate abstractions;
- Apply efficient algorithms to the processed data;
- Finally, build systems that scale.

But above all, human agency is essential. Algorithms and systems don't carry intrinsic value. They only become useful when they empower people and enhance society. Technology is not inherently good or bad, it's how we choose to use it that defines its impact. In 2021, Meta developed smart glasses in partnership with Ray-Ban. The technology was very impressive. The glasses allowed users to take photos and videos, make calls, and listen to music without needing to pull out their phones. The pitch was "hands-free convenience" and seamless integration with your digital life. However, despite the technological advancement, people felt uncomfortable being around others wearing the glasses and some establishments even banned them outright. Rather than feeling empowered, users and bystanders felt vulnerable.[4] The opposite also happens. When the first AI models that power ChatGPT were developed, even the researchers working on them were astonished by their broad applicability to tasks they didn't have in mind to begin with. The technology proved to be very empowering in many scenarios, like answering complex questions about research publications, and it still amazes some of its early inventors. Through the book, I'll discuss several other examples of technologies that either positively or negatively impact society.

The reason I cover so much ground, starting from bits and moving all the way to complex systems such as AI and quantum, is to build our collective understanding of technologies and to develop our human agency. Let's consider sustainability. By covering how algorithms work, how they can be used in complex systems, and how cloud computing primitives support running these services at scale, we can have an informed discussion of how computational resources are used to run services we know and love, such as Google Search or Amazon.com. As we'll see in Chapter 4, efficient algorithms can lead to better resource utilization, minimizing the carbon footprint of services, and positively impacting sustainability. We'll

---

4. https://www.business-humanrights.org/en/latest-news/ray-ban-meta-glasses-raise-concerns-about-privacy-rights-consent-and-increased-surveillance (Last accessed in July 2025).

understand how efficiency metrics such as resource utilization and power usage effectiveness (PUE) can help us have an informed discussion about sustainability and the impact of our power-hungry cloud computing industry in the world.

The last topics, AI and quantum computing, are the most complex, but by then we will have the tools to understand them more deeply. Technical readers will notice that I opted for clarity of presentation over completeness of some concepts. For instance, when talking about web search, I omit the use of caches, which are systems that store previously seen queries and their results to increase the overall efficiency by avoiding processing the same popular user queries over and over. I intentionally opted to go over more subjects at a shallower level, as opposed to conveying a smaller number of areas at a deeper level. Interested readers will find plenty of resources to dive deep into algorithms and systems they find interesting. I tried, as much as possible, to present technologies in chronological order, from the early days of the internet to today.

Chapter 5, on organizational efficiency, doesn't describe a new technology. In that chapter, instead of computer systems, I discuss how to run organizations to produce the most value, at the lowest possible cost, by building on and incorporating the concepts outlined in the rest of the book. My motivation to include this chapter was twofold. First, it provides a systematic approach for the organizational productive cycle. Innovations that generate efficiencies, positively impacting the bottom-line, allow us to reinvest our gains on innovations that produce value and increase the top-line, generating a virtuous cycle for organizations. Second, it gives us a good foundation to talk about the effects of AI in productivity. Perhaps one of the most foundational questions we need to answer as a society is how will the new AI systems affect the way we work. And given that, what changes should be done to the educational system to prepare for such a workforce. Chapter 5 provides a foundation for that discussion.

Writing this book has been one of the most interesting projects of my professional career. As Richard Feynman said, it is challenging to present complex concepts at the freshman level. And although I know that no one can become an expert in computer science by reading a book, hopefully people will find the topics presented here interesting regardless of their level of technical knowledge. Most importantly, I hope the book is fun and that it helps broaden our collective understanding about technology, and how it can have a positive effect in our lives.

In order to help understand some of the technical concepts presented here, I built an accompanying website, https://digitalagencybook.org, that provides extra material and shows visualizations for some of the algorithms I described throughout the book. Finally, all author proceeds from this book are donated through Microsoft Philanthropies to causes related to the democratization of computer science education.

1.

Basic concepts

Roger Federer, efficiency, bogosort, antifragility, self-driving cars, and digital-only myths

> "Why don't we just shoot the sumbitch?"
> Harrison Ford

The swordsman displayed impressive dexterity and was ready to fight. But instead of engaging, Indiana Jones just shot him, ending the duel instantaneously. This iconic scene from the 1981 film *Raiders of the Lost Ark* is still considered one of the funniest scenes from the Indiana Jones franchise. A fun fact is that it was an improvisation by Harrison Ford. Ford explained "I was sick, and besides, up to that point, I kept worrying about the fact that I had been wearing this gun that I had never drawn. So, I said to Steven [Spielberg], 'Why don't we just shoot the sumbitch?' He said, 'Okay.' He was ready to get out of there too. That's how we got that scene."[1]

Efficiency often reveals itself through contrast. Think of that famous Indiana Jones scene. The swordsman dazzles with grace and flourish, but Indiana just pulls out a

---

1. https://www.slashfilm.com/1484507/indiana-jones-harrison-ford-swordsman-shoot-disturbed-writer/ (Last accessed January 2025).

gun. Decisive, effective, and efficient. It's funny precisely because it subverts our expectations of effort. Aristotle once defined comedy as a portrayal of people "worse than average," out of sync with what's expected. I feel that every time I try to play tennis, my coordination being clearly below the comedic threshold. Compare that to Roger Federer, whose effortless strokes embody strategic minimalism. He conserves energy not by doing less, but by eliminating the unnecessary. As a result, there is nothing comedic about his technique. Organizations work the same way. A lean team, like Federer on the court, achieves harmony by shedding excess. Too many people in a meeting is like swinging five rackets at once. It's not just inefficient. That disharmony, as Aristotle noted, becomes its own form of comedy.

The U.S. faces a substantial $36 trillion debt as of January 2025. While opinions on the severity of this deficit differ, we can agree on the importance of investing in efficient projects. These are projects that produce the desired results without wasting resources, like infrastructure projects that finish on schedule and under the forecasted budget, or organizations that produce value without being overstaffed. This is easier said than done, as the natural state of things gravitates towards chaos and disharmony. Left unchecked, projects would stall, companies and cities would grow in a disorganized way, and energetic resources would be wasted.

As we talk about computer systems throughout this book, efficiency will be a recurring theme. As it is a broad concept that may vary between contexts, I do not attempt to provide a formal, singular, definition that can be applied to every context. Rather, I'll provide examples demonstrating varying definitions in several domains, highlighting the common unifying concepts that will compose an efficiency framework. My goal in developing this framework is to enable us to view the world through the lens of a computer scientist. Why? Computer scientists are trained to build efficient systems. Implementing these efficiencies will enable us to better manage resources, allowing us to scale our systems and generate value. Efficient systems are cheaper to operate and allow us to reinvest the gains into projects that will positively impact society. In the rest of this chapter I'll introduce some basic concepts that will be used throughout the book, including the concept of myths, and some properties of systems, such as fragility and antifragility.

Algorithmic efficiency

Computer scientists learn about efficiency early on, in their CS101 class. Although this is not a traditional computer science textbook, let's start with algorithmic efficiency. Both to demystify it and because it is a foundational concept used throughout the book. Algorithms are nothing more than recipes that instruct the computer to solve problems. One of the first algorithms we learn in CS101 is sorting. Given a list of numbers, let's say 31, 27, 101, 1, -1, and 0, a sorting algorithm returns the sorted list containing the same elements from the original list. In this case, a correct sorting algorithm would return -1, 0, 1, 27, 31, and 101 if sorting in ascending order, or the reverse list otherwise.

The algorithms for sorting in ascending or descending order are basically the same, so let's just focus on ascending order for now. Solving a sorting, or any problem, with a computer means thinking about a recipe that the computer can execute. This recipe is the algorithm, which is described in a language, called the programming language. Sorting a list in ascending order requires comparing each number with every other number in the list. There are many sorting algorithms that can do the job, each with slight variations and tradeoffs. Similarly, in our daily lives, some people sort their clothes by spreading all of them in the bed before rearranging them in a single pile by color, while others sort by splitting clothes into a few piles. One of the simplest sorting algorithms is bubble sort. Its name is an indication of how it works. It "bubbles" up the largest elements to the top.

Bubble sort starts with the unsorted list and compares the first element with every other element in the list. After each comparison, the largest of the two elements is "bubbled" up, so that after this step the largest one would be in the last position. If we compare 31 with 27, as 31 is larger we'd swap them. The list would then be 27, 31, 101, 1, -1, and 0. Then we would compare 31 with 101 and keep the list intact. In the next three steps we'd compare

101 with 1, -1, and 0 and swap every time, arriving at the end of this first step with the modified list 27, 31, 1, -1, 0, and 101.

1st pass

31,	**27,**	101,	1,	−1,	0
27,	**31,**	**101,**	1,	−1,	0
27,	31,	**101,**	**1,**	−1,	0
27,	31,	1,	**101,**	**−1,**	0
27,	31,	1,	−1,	**101,**	**0**
27,	31,	1,	−1,	0,	101

2nd pass

27,	**31,**	1,	−1,	0,	101
27,	**31,**	**1,**	−1,	0,	101
27,	1,	**31,**	**−1,**	0,	101
27,	1,	−1,	**31,**	**0,**	101
27,	1,	−1,	0,	31,	101

3rd pass

27,	**1,**	−1,	0,	31,	101
1,	**27,**	**−1,**	o,	31,	101
1,	−1,	**27,**	**0,**	31,	101
1,	−1,	0,	27,	31,	101

4th pass

1,	**−1,**	0,	27,	31,	101
−1,	**1,**	**0,**	27,	31,	101
−1,	0,	1,	27,	31,	101

5th pass

−1,	**0,**	1,	27,	31,	101
−1,	0,	1,	27,	31,	101

Figure 1. All the bubble sort passes required to sort the full list. The comparisons in each step are highlighted. During each pass, we "bubble" the next largest element towards the end of the list. In this case we just need 5 passes to sort the full list, as the list was partially sorted with -1 and 0 in the correct relative order.

If we repeat this recipe a few more times we will end up with a sorted list. In this example, as we have six elements in the list, we'd need to do a maximum of six passes. In a single pass, we compare each element with every other element, which in this case amounts to six comparisons per pass. This results in *6 × 6* comparisons across the total six passes. Figure 1 provides a visual illustration.[2] This number of comparisons is a proxy for how many instructions the computer would need to execute to fully sort the list. The computer would therefore need to execute 36 comparisons and swaps for a list of size 6, and n^2 comparisons and swaps for a list of size n. The algorithmic efficiency of the bubble sort algorithm is therefore n^2. Throughout this book I'll introduce several measures of efficiency. The first one is the **order**, which is a proxy of the maximum number of instructions an algorithm would have to perform to solve a problem. The order for bubble sort is n^2, commonly represented as $O(n^2)$, which is referred to as the **big-O notation** for measuring algorithmic efficiency.

A computer's speed is measured by its processor's frequency. A typical laptop or mobile phone in 2025 has 3GHz frequency, which means 3 billion cycles per second. Considering for simplicity that we can perform one instruction per cycle, in one second these processors would be able to perform 3 billion swaps and comparisons for our bubble sort algorithm. The good news is that, unlike humans, computers tirelessly repeat the same instructions over and over without complaining. If we are willing to wait one second to sort our list, we'd be able to sort a list of at most 54,000 elements with bubble sort.

This is very good. However, despite being simple to explain, bubble sort is not the most efficient of the sorting

---

2. Please see a visualization of bubble sort at https://digitalagencybook.org/visualizations/bubble-sort.

algorithms. If we use a faster algorithm instead, such as quicksort, we'd be able to sort a list of approximately 112 million elements if we waited the same one second using the same 3GHz processor. That is a much bigger list, about 2000 times bigger. In other words, quicksort is 2000 times more efficient than bubble sort. With a more efficient algorithm we can sort the list with less instructions, which saves time and energy. In terms of efficiency, the worst thing one can do is leave a computer idle, consuming energy without producing anything useful. Arguably, other bad options are spending the day browsing social media, draining computational resources without producing useful outputs, and running inefficient sorting algorithms.

But don't worry, it can always get worse. Instead of bubble sort, which is relatively efficient, we could instead use bogosort, also known as stupid sort. Its recipe is simple:

1. Shuffle the list arbitrarily.
2. Check if the list is sorted.
3. If not, repeat step 1.

Anyone who ever needed to sort anything knows this cannot be a good idea. As we've seen before, inefficiency is funny. On average, in the same 3GHz device we'd be able to sort a list of at most 12 elements in one second using bogosort. For a list of size n there are $n!$, or n

factorial,[3] possible permutations. Since Bogosort doesn't check these permutations in any particular order, there is no way to guarantee it will ever finish. In the worst case it could take infinite time, if it is unlucky and never shuffles the list in the correct sorted order. On average, however, it'll likely finish after $O(n!)$ tries, requiring a lot more instructions than our $O(n^2)$ bubble sort.[4]

It is clear by now that the choice of the algorithm is a key factor for sorting, but you may be wondering why sorting is a relevant problem to begin with. If you are, for instance, looking for restaurants that have the top reviews in your area you need sorting. To organize your emails with the most recent ones showing on top, you also need sorting. The same is true for many other applications in commerce, finance, logistics, and management, just to name a few. Computers spend a lot of their cycles sorting, but as I mentioned before, they don't mind doing repetitive work.

Even in AI applications, all that computers know how to do is to relentlessly repeat instructions, obediently following the steps described in the algorithms we ask them to execute. Besides being obedient, computers also possess the desirable qualities of speed and accuracy.

---

3. The factorial of n, is $n! = n \times (n\text{-}1) \times (n\text{-}2) \times ... \times 3 \times 2 \times 1$. For large values of n, the factorial is larger than n^2 and even larger than 2^n.

4. Please see a visualization that compares bubble sort and bogosort at https://digitalagencybook.org/visualizations/bubble-vs-bogo.

They perform instructions much faster than humans, and with much fewer mistakes. AI algorithms are impressive, and have the potential to revolutionize society. But if you look under the covers, the computers are simply following orders to execute simple instructions, such as orders to multiply numbers and matrices.

In fact, even if you could develop an algorithm to consume the computational resources of every processor in the world, in data centers, homes, cars, microwaves, and offices at the same time, this super computer program would be computing a function that takes an integer number as input, possibly a very large one, and returns another integer as the result. This is true for everything that computers do. Computers can only compute functions mapping integers to other integers.

The simplest way to understand this is to visualize that the inputs to algorithms and their results must be represented in computer memory. In this simplest form, a processor reads the memory, performs some instructions, and writes back to the memory. Computer memory is just a collection of **b**inary dig**its**, or **bits**, which are the smallest unit of information in a digital system. Since bits are either 0 or a 1, they can be read as integers. After 0 and 1, bits 10 in binary represent number two, 11 number three, 100 number four, and so on. Even if we combine all the memory of every computer in the word, it will still be just a longer sequence of bits. If the algorithm writes a file or displays something on the screen, still the contents of the file and the screen are just extensions of this large bit sequence. At the end, all computers can do is to perform instructions that manipulate numbers, represented as sequences of bits.

But wait, we just saw that computers can sort lists, returning an organized list of restaurants with the most positive reviews. We also know that they perform complex AI tasks, such as producing a historically accurate text about the Roman Empire. How can they do that if all they know how to do is compute integer functions? Welcome to the world of abstractions.

The magical world of abstractions

While it is true that computers only know how to manipulate sequences of bits, we are free to slice and dice and interpret these bits in many different ways. One common way is to represent numbers using 32 bits. The choice of 32 bits is historical and was popularized by Microsoft's Windows NT in the 90s, as computers used to be able to process only 32 bits in a single instruction. If we use 32 bits per integer, we'd need 192 bits to represent the six numbers in the input list from our bubble sort example, and another 192 bits for the result. As a typical memory size for a laptop computer is 16GB, or 128 billion bits, we would have no problem representing very large lists.

It is up to the algorithm developers to decide what the sequences of bits represent. This is done through **abstractions**, which can be anything from individual numbers, to lists, matrices, websites, songs, videos, and the algorithms themselves. Programming languages allow programs to not only describe the computation that needs to be done, but also represent the inputs and outputs and every piece of information that needs to be processed using abstractions. Abstractions are powerful because they allow developers to think about lists, songs, and websites, instead of sequences of bits.

A banking application must represent its customers, who must have physical addresses, passwords, and a list of accounts. Each account has a balance and a list of transactions. Each transaction has a value and source and target accounts. This gets very complicated very quickly, and it is up to the algorithm developer to come up with these representations. Two different banking applications can be implemented using completely different representations for customers and completely different algorithms for processing transfers.

Our human-designed computer systems bear profound similarities with the world we humans inhabit. In *Sapiens*,[5] Yuval Noah Harari highlights the differences between biology and culture. Biology evolves slowly, while culture evolves much faster. It is much easier to transform society by inventing a new **myth**, such as the importance of social mobility, than to wait for evolutionary changes to alter our DNA. Language allows us to define myths such as countries, companies, money, and social mobility. These complex concepts which we

5. *Sapiens: A Brief History of Humankind*, Yuval Noah Harari, Harper, 2015.

simplify and understand as myths, even if they feel to be true, have no real representation in the real world, which is governed by biology, chemistry, and physics. A similar divide exists for computer systems.

Computer hardware is physical. The term "cloud computing" may sound ethereal but it is a combination of buildings full of computers, network cables, batteries, storage disks, and several other components, which are all very much subjected to the **laws of physics**. Cloud computing also has computer software, which are algorithms implementing the cloud computing abstractions that allow you to run your services and backup your phone. Like our DNA, hardware evolves slower than software. It is true that processors get incredibly faster and memory gets bigger over time, just like humans get fitter with evolution. The famous Moore's Law states that the number of transistors in computer chips roughly doubles every two years, leading to exponential increase in computing power and efficiency. Yet, the physical design of computers today still follows the same model proposed by mathematician Alan Turing in 1936.

As computer systems can only compute integer functions, it is up to humans to interpret what these numbers mean. These representations are abstractions that allow us to model the complex concepts from our world using the simple computer language of zeros and ones. Like culture, abstractions evolve much faster than computer hardware. New computer games, dating apps, and supply management systems are produced daily. There are, however, important differences between computer abstractions and what Harari defines as myths: computer abstractions can be used to represent either physical or digital-only concepts.

One of the first applications built on computers were numerical computations to support the development of the atomic bomb during the Manhattan Project. The atomic bomb is very real and not a myth. On the other hand, although humans are social creatures and have interacted in complex social structures for centuries, social networking platforms are completely digital and have no counterpart in the physical world. If social networking platforms were outlawed, the world would continue without major disruptions.

There have been a couple of times in my career as a corporate executive when I had to explain that if we change our employee performance review system the world would not collapse. It is shocking to some, but not requiring a 360 degree review of a particular employee at their annual review is different than eliminating gravity. The world doesn't collapse and everyone will still be reviewed and bonuses will be paid. We become so accustomed to the myths that form our culture that we forget that they can be changed. However, changing the physical world is much harder. Once we decide to build a road with two lanes and not four, it is very hard to make the expansion once we discover that the surrounding cities grew substantially and traffic is now unmanageable.

Laws of physics and laws of fiction

Nassim Nicholas Taleb describes three distinct types of systems in his book *Antifragile:*[6] fragile, robust, and antifragile. A fragile system is one that breaks easily, like a porcelain plate. A robust system is one that does not break easily, such as a well built bridge. However, as Taleb explains, robustness is not the opposite of fragility, since a robust system does not get stronger when stressed. A bridge does not get safer if heavily used. Antifragility is the true opposite of fragility. Antifragile systems become stronger in the presence of stressors. Evolutionary biology is an example of an antifragile system, as over time the species will typically select beneficial traits in the face of adversity. Future generations evolve to better face such adversities of the past.

Efficiency only exists in robust and antifragile systems. A bridge made from cheap materials may appear to be deceivingly inexpensive, but it is not actually cost efficient in the long run. A fragile bridge, or any fragile system, is extremely inefficient as it will require ongoing maintenance and have to be rebuilt should it collapse, not to mention all the suffering and disruptions such a collapse would cause. For a system to be considered efficient, it must be at least robust. Being antifragile would be ideal, in scenarios where that is possible. Despite all his grace, Roger Federer would be a fragile system if he didn't

---

6. *Antifragile: Things That Gain From Disorder,* Nassim Nicholas Taleb, Random House, 2012.

know how to hit the ball, similar to an elegant and concise sorting algorithm that fails to produce the correct output.

An efficient bridge has to be properly designed to begin with, starting with the blueprints, the design and building process, and using the right quantity and quality of each material. The rigidity of the real world has a silver lining. The laws of physics are well understood and physical objects are much less prone to **black swan** events:[7] unexpected events that may be catastrophic. It is very unlikely that one morning, several trucks weighing 20 thousand tons each, which is a thousand times heavier than the average truck, will simultaneously go over a bridge causing it to collapse. A computer search system such as Bing.com, however, may see an increase in traffic of a several thousand fold when a global celebrity, like Michael Jackson, dies. Although celebrities dying is an expected natural event, which we'd not classify as a black swan event, Michael Jackson's unexpected death was a black swan event for search engines in 2009, which weren't accustomed to such a sharp increase in query volume.

When designing computer systems, we must expect these black swan events will happen. We shouldn't assume that all swans are white just because we've never seen a black one. Similarly, we should not assume that Bing.com traffic will never increase several thousand fold just

---

7. *The Black Swan: The Impact of the Highly Improbable,* Nassim Nicholas Taleb, Random House, 2012.

because it hasn't done so in the past. This is fundamental to the design of computer systems, and this is why building robust or antifragile systems is challenging. This is a high price to pay, but you have the advantage of not being constrained by the rigid constraints of the real world. A search system can always be modified as we see fit, as it has no physical counterpart. I jokingly like to say that computer systems are subject to the **laws of fiction** instead of the laws of physics.

Just like search engines, cities are myths not governed by the laws of physics. However, cities are not just a concept, as they are supported by infrastructure and contain buildings, such as the city hall and the fire stations, and are governed by municipal organizations, such as the mayor's office and the police. They also house computer abstractions like their employee databases and the tax audit systems. With so many real-word interdependencies, dismantling the myth of cities may prove to be even harder than redesigning a bridge.

We can distinguish between two types of myths: those anchored in physical reality and those sustained purely by a shared belief in the digital realm. Cities are physically-backed myths. They are not just collections of buildings, but embodiments of centuries of rituals, governance, and legal fictions that are deeply embedded in our lives, history, and cultures. By contrast, social networking platforms, like TikTok, are digital-only myths. They rely on collective belief in their relevance and exist primarily in the virtual world. Physically-backed myths are harder to dismantle not because they are more "real," but because they are reinforced by institutions, infrastructure, and long-standing systems that give them enduring power. The true difference is how deeply and durably they are woven into the fabric of society.

For example, due to ongoing user data concerns and non-US ownership of the platform, in January 2025, TikTok was briefly banned in the U.S., highlighting how quickly a digital myth can lose its hold, with little lasting disruption to broader societal function. Users may switch to another platform, if one is available, but overtime, digital-only myths can be dismantled. Yahoo and Skype, which were pioneers in web portals and videoconferencing, respectively, are examples of digital-only myths that once were so prevalent in our lives and are now forgotten. Although we should question physically-backed myths, and we'll do that throughout this book, improving their efficiency or dismantling them would require major effort, especially when compared to digital-only myths.

The term digital-only is not completely accurate. The computer abstractions are physically stored in computer hardware, and need energy and computational resources to do anything useful. This is akin to Harari's myths, which despite not being governed by physics, still require our brain power, energy, and communication skills to be useful. In the absence of a more precise term, I'll stick with "digital-only" to convey this type of myth throughout this book.

Before we move on, we can start defining some taxonomy. Systems can be **physical**, such as bridges, **physically-backed myths**, such as cities, or **digital-only myths**, such as social network systems. Physical systems are governed by physics, biology, and chemistry. They are harder to change but less prone to black swan events. Digital-only myths are governed by fiction. Easier to change but also more susceptible to black swan events. Physically-backed systems are governed by a combination of physics and fiction. They suffer from the two problems, being hard to change and still susceptible to black swan events.

These three types of systems can be **fragile, robust,** or **antifragile,** but only robust or antifragile systems can be efficient. It is a futile effort to analyze the efficiency of a sorting algorithm that doesn't sort, or a bridge that has already collapsed. Bogosort is extremely inefficient, but its implementation can be robust if it always produces the right result. Physical systems and physically-backed myths may benefit from computer **abstractions**, while digital-only systems only exist as computer abstractions.

The physical system "car"

We can pretty much all agree that a car is governed by the laws of physics. However, more and more car components are being augmented by software over time. In electric vehicles, the motor and braking systems are software-controlled for smoothness and for storing energy during deceleration. Apart from these physical components that are enhanced through computer abstractions, cars also have digital-only components, such as music systems, sensor controls, and self-driving features.

Tesla cars have so many digital-only components that their recalls happen as over-the-air updates, similar to how mobile phones are updated. That feature alone greatly increases the efficiency of maintenance, which except for flat tires and cracked windshields, happens almost invisibly. The process of transforming a physical system, such as car maintenance, into a digital-only one is sometimes referred to as **digital transformation**. This term is hyped and much talked about because it normally comes with efficiency benefits. For example, digital braking systems are easier to update, online newspapers save trees and are easier to consume, and digital photos don't need to be developed.

The most interesting physical systems, such as cars, are composed of subsystems which can be either physical or digital. In the absence of teleportation and the dystopian scenario in which our lives become completely digital, cars will continue to be physical systems, as they have the required physical function of transporting humans. However, the digital transformation of car subsystems has the potential to make transportation much more efficient.

There is an automation level classification for self-driving cars. Level zero means no automation at all. Level five means full automation, with no driver needed and without geographic constraints. Currently, as of August 2025, there are no level five self-driving cars in the U.S. and the only level four system in production is Waymo's Robotaxi service. Level four means full autonomy, but only in specific conditions and geofenced areas. The car must safely stop under adverse conditions. The Waymo service is now operating in Phoenix, San Francisco, Austin, and Atlanta.[8]

The popularization of services such as Robotaxi would enable people to switch from owning a car to joining a subscription service that works just like ride-sharing services, such as Uber or Lyft, without the driver. Whenever you want to go somewhere you can request a ride and the service would dispatch the car closest to you to take you there. Without the need for drivers, services like Robotaxi will be both safer and cheaper than ride-sharing services. Self-driving cars and Robotaxi services are both examples of antifragile systems.

The self-driving feature learns from data, so the more people there are using the service, the better it gets. The more cars drive on an unknown route, the more details from that route could be recorded. The more data is collected on how other cars react to difficult situations, the better they can estimate blind spots of other cars. If accidents happen, their data could serve as valuable information as well. For our typical level zero car, with no automation at all, if an accident happens, the lessons are usually limited to the involved parties. They are the only ones who can learn anything from

---

8. https://www.wsj.com/tech/uber-lyft-self-driving-taxis-a3659c9c (Last accessed in January 2025).

what caused the accident and how it could have been prevented. Unknown routes, difficult situations, and accidents are all stressors that make the self-driving system better over time.

Robotaxi services also benefit from stressors. A popular event could cause spikes in the service utilization that would, in turn, lead to better provisioning and scheduling algorithms for the fleet. Over time, these combined improvements would generate even more efficiencies: with a larger number of safer self-driving vehicles they would be able to drive closer together, reducing congestion. With widespread adoption, there likely would be less parked cars at any given time, decreasing the need for parking lots. Parking lots could be converted into efficient spaces, such as housing and green areas, and we could have major changes in the design of urban spaces. We could even fulfill the prophecy that American garages eventually become storage space without much remorse.

Every year there are about 40 thousand deaths in the U.S. due to motor vehicle accidents, which is about 1.3-1.5% of all deaths in the country. In addition, over 80% of new cars sold between 2022 and 2024 were financed, and currently more than seven million Americans are three months behind on car payments. This data may change over time, but the point is clear. There are several efficiencies that can be gained if there were a widespread deployment and adoption for services like Robotaxi, assuming the technology will get cheaper over time and it will be properly regulated.[9]

We started talking about a physical system, the car, and how the digital transformation of some of its subsystems could have widespread impact. It is hard to precisely quantify how much efficiency could be gained, as there are so many variables and the technology is still evolving, but they would include gains in many areas, from accident prevention to energy efficiency and to urbanism. The early evidence is very positive. Waymo analyzed 7.13 million miles of fully autonomous driving in Phoenix, Los Angeles, and San Francisco. The analysis shows a reduction of 85% in injury-causing accidents when compared to human drives.[10] However, there is no free lunch. Digital transformation means replacing a physical system with a digital-only myth, making it susceptible to black swan events.

In the case of self-driving cars, black swan events would include systemic failures. It would be hard to get every driver in New York City simultaneously drunk, so accidents caused by drunk drivers are typically isolated events. A systemic failure, either caused by an infrastructure failure or by malicious hackers, could theoretically cause cars to behave as if their drivers were all

9. https://www.nber.org/system/files/working_papers/w24349/w24349.pdf (Last accessed in June 2025).
10. https://www.businessinsider.com/waymo-driverless-cars-data-safer-than-human-driven-vehicles-2023-12 (Last accessed in June 2025).

simultaneously drunk, causing major damage. We have to assume that black swan events will eventually happen when building systems. However, I used the word "theoretically" here as, in this scenario, there are enough checks and balances in the system that could prevent such a catastrophe from happening. As I mentioned before, the self-driving system is antifragile and gets better in the presence of black swan events. These checks and balances have been developed by learnings from prior black swan events, so generalized systems failures become more difficult to happen over time.

Minimizing the impact of black swan events is essential, especially as a growing part of our lives is built on digital abstractions. Unfortunately, not all digital-only systems are antifragile. Many are quite brittle when stretched beyond expected conditions. This is a recurring theme throughout the book. For now, consider this: if a widespread outage were to knock out the internet, many homes would lose the ability to unlock doors, control heating, or water the plants, given that the apps that control these systems would be offline, with "smart homes" being more susceptible to black swan events than traditional homes that do not integrate these features. In such cases, having a manual override, such as an extra pair of keys, can make all the difference. The lesson is simple. For every elegant abstraction, we should maintain a grounded, physical alternative. Resilience and antifragility do not come naturally from innovation, they must be engineered.

Despite the gains, there may also be societal impacts of self-driving cars, as highlighted by a recent backlash against the perceived Waymo surveillance.[11] Any new technology requires a broad societal debate and careful adoption, making sure it produces more value to the society than causes harm, and its negative aspects are minimized. As exhaustively discussed in *Power and Progress,* technology per se is neither enriching nor predatory, but its use can be either. Self-driving cars and Robotaxi are no different. We need human agency and broad societal support to make sure their broad adoption is successful.

The physically-backed myth "Amazon.com"

Amazon.com is almost synonymous with efficiency. Sort products by their review and purchase them with a single click. It started over thirty years ago with the internet commerce boom. The initial focus on books was a careful choice, which proved to be the right one. Books are non-perishable and uniform, meaning that every single copy of a book is exactly the same. This makes books attractive from the retailer viewpoint, as they can be stored indefinitely and don't need to be inspected for quality assurance. However, two other factors were even more crucial for Amazon's success: global reach and catalog completeness.

---

11. https://www.theverge.com/google-waymo/682932/la-protest-waymo-fire-destroy-ice-police (Last accessed in June 2025).

The market for books is global. Books published in one country can be sold everywhere in the world. This means that an online store has the potential to reach a lot more of these global customers than any of the existing brick-and-mortar bookstores. This is again a digital transformation, from physical stores to a digital-only store. In addition to all the efficiencies that come with it, such as not paying rent and not having to hire cashiers, the customer base is amplified from the store's neighborhood to any neighborhood that has an internet connection and that does not fall within business or shipping constraints.

The last factor that made books the right choice for Amazon is that the book market already had a small number of large and efficient distributors, such as Ingram and Baker & Taylor. This allowed Amazon to have a large catalog without having to deal with several small publishers. The Amazon book catalog was not only large, it was exhaustive. There is virtually no published book that was not part of its catalog. This sense of completeness is a key differentiator and the same reason that Netflix and other streaming services eventually displaced physical video rental stores, such as Blockbuster.

In the research paper *Anatomy of the Long Tail: Ordinary People with Extraordinary Tastes*[12] a group of

---

12. Anatomy of the Long Tail: Ordinary People with Extraordinary Tastes, Sharad Goel and others, WSDM '10: Proceedings of the third ACM international conference on Web search and data mining.

Yahoo researchers showed that most people like a combination of popular products and less popular peculiarities. It is certainly true for me. I'll read all the popular Malcolm Gladwell and Sally Rooney books, but also much less popular books about science, including some very obscure ones about computer science. Their paper debunked the theory that most people have popular tastes while others form an eccentric minority. Everyone has a unique taste, and as a result, the complete catalog transforms online retailers into a one-stop shop for everyone.

Amazon has since expanded from books to almost everything imaginable, but its focus on efficiency remains. I visited one of Amazon's fulfillment centers in 2024. It is a giant warehouse that stocks all of the products that Amazon sells. When given a customer's order, computer-controlled conveyor belts route products from all over the warehouse to be packed, labeled, and dispatched in trucks. The goal of the warehouse operations is to minimize the time to fulfill each order while maintaining a minimal error rate. Some aspects of the process are manual, but most are done by a combination of robots and software. Of course, other companies also have efficient software controlled warehouses, such as Kroger and Walmart, with their partnerships with Microsoft.

During my Amazon warehouse visit I saw a robot that minimizes the amount of cardboard used for packing single product offers. The robot was programmed to build customized packages for each product, given its dimensions, simultaneously building the package and packaging the item. There are several checks to ensure that Amazon's fulfillment system is robust. As each package goes out, for instance, its total weight is compared with the sum of the specified weights of each product that is supposed to be in the order, using a scale like those in the bagging section of a self checkout system at a grocery store. If the sum is off, the package is set aside for inspection. Amazon's efficiency in its retail business comes from process and algorithmic optimizations throughout the entire system, from purchasing, to the warehouse, to delivery, with attention to detail at each step.

Another fact surprised me during my visit to the warehouse. I saw relatively few books compared to other products. When I asked our host about that he said: "we invented the Kindle." Although an optimized supply chain greatly reduces cost, completely eliminating books through digital transformation is even better. Electronic books have negligible costs and no downside from the retailer's viewpoint. They also have advantages for readers. Kindle allows you to continue reading when the airplane lights are off with no impact to your neighbours. You can also see relevant passages highlighted by other users. It works when disconnected from the internet and, if you have misbehaving pets and children, it may be even more robust than physical books, as the Kindle is less likely to be eaten or torn apart. This type of digital transformation generates not only efficiencies, but also provides added value to consumers, as the digital version of the object has more bells and whistles than the physical one. However, digital books don't have all the benefits of a physical book, so digital books are not a replacement. I, for one, very much enjoy holding a physical book and use a combination of both formats. Human agency means not embracing new technologies simply because they are more efficient, but also considering human factors, which might include enjoying the scent of opening the pages of a new book.

Despite having warehouses full of products, offices, and more than one million employees, Amazon.com is still a myth, according to Yuval Noah Harari's definition: the concept of Amazon.com does not exist in the physical world and it is not governed by the laws of physics and chemistry. Should it cease to exist tomorrow, its physical assets would simply be repurposed, and life would continue. Amazon is an efficient myth. It generates value by consuming resources in an optimized way. Being an efficient-minded company, our expectation is that it will over time improve its efficiency even more, either through organizational changes, procedural and algorithmic optimizations, or other, still to be invented, digital transformations. These efficiencies allow the company to provide cheaper services and to reinvest in new technologies that will eventually become novel product lines.

The digital-only myth "spreadsheet"

Over four billion users worldwide use either Microsoft Office or Google Workplace. This is almost one license per living adult. If spreadsheets were outlawed tomorrow, several knowledge workers such as product managers and accountants would be less productive and a few tech employees who work on these projects would lose their jobs, but most of the world would remain intact. Spreadsheets are a digital-only abstraction, with no physical counterpart.

Spreadsheets on their own do not pose harm to anyone and the chance of them being outlawed is very slim. It is more likely that instead of being outlawed, they could be replaced by a newer, simpler to use technology. And this is why tech companies' valuations are so high. The top ten companies by market capitalization are all tech companies. No one would be surprised if a new startup invents something more intuitive and better than spreadsheets or if Google or Microsoft comes up with new services as popular as their productivity tools.

Digital-only myths don't necessarily imply low-cost or efficiency. When I asked ChatGPT "How much does it cost to train the model you are using to answer this prompt?," it responded that the model being used was GPT-4 and, although exact numbers are proprietary, estimates were in the $50-100 million range. This is a 4-8 times increase from the previous GPT-3. Despite using efficient algorithms, the costs are extremely high. Running Microsoft Office resiliently worldwide is both costly and complex: the service must be always available, multiple concurrent modifications to documents must be processed without failures, and user data may never be lost, just to name a few hurdles.

Besides being expensive to operate, building some of these systems may take time. Several of the platforms that we use in our daily lives, such as the Windows operating system, Google search, and our banking and health care systems, took years to be developed. Engineering teams continuously work on new features, efficiency improvements, and antifragility enhancements to keep up with new demands from users and to maintain the services operating well.

Humans have used tables to organize information since antiquity. Spreadsheets are the result of their digital transformation. There are digital-only myths, however, that have no parallel in the physical world. As Google's chief technologist Prabhakar Raghavan likes to say: "before Twitter, we never knew humans had the necessity to tweet." Before social networks, humans didn't Snap and didn't count likes either. These are digital-only behaviors that only have meaning in the digital world.

In 2022, Curzio Research bought 19 properties for five million dollars.[13] Depending on where you're from, this might sound like a fantastic deal. However, what Curzio Research purchased was not actual physical land, but rather digital tokens that represented the ownership of the properties. These non-fungible tokens, or NFTs, are digital-only myths. Non-fungible just means they are

---

13. https://www.financemagnates.com/thought-leadership/tcg-world-announces-50m-purchase-of-virtual-estate-in-the-largest-open/ (Last accessed in January 2025).

unique tokens. In Curzio Research's case, these tokens uniquely represented physical real estate in some virtual world. As far as I can tell, there are an infinite number of possible virtual worlds, so I don't know how the supply and demand curve applies for virtual real estate. NFTs can be anything. The first tweet ever posted was "'just setting up my twttr' by Jack Dorsey, Twitter cofounder and former CEO. In March 2021, it was turned into an NFT and sold for $2.9 million dollars. The buyer then attempted to resell it in an auction a year later for $48 million, claiming part of the proceeds would be donated to charity. The highest bid was only $280 dollars[14].

It appears that NFTs for tweets, digital art, and real estate were just a fad, but the virtual world is here to stay, and I hope this book also helps us rationalize what myths and abstractions are improving our human experience, making us happier, more productive, and smarter, and which ones are detracting from our humanity.

An efficiency-based outlook

I don't believe in a relentless search for efficiency. A world in which every organization had the perfect number of employees, building highly-optimized software-controlled conveyor belts and sorting lists with quicksort, would probably be very bleak. Going back to Aristotle's

---

14. https://www.forbes.com/sites/jeffkauflin/2022/04/14/why-jack-dorseys-first-tweet-nft-plummeted-99-in-value-in-a-year/ (Last accessed in January 2025).

definition of comedy, a highly efficient world wouldn't be very fun. We need to walk on the longer, scenic route, to have pointless conversations, and also read printed books despite knowing that the Kindle is more efficient and better for the environment. Human connection is important, and it may be the reason for some techno-logical devolutions, such as the resurgence of brick-and-mortar bookstores.[15]

The innovation process is also inherently inefficient. New ideas come from trial and error. And as every startup investor knows, most new ideas fail. Another way to look at this: if everyone is already busy executing a task, like in an assembly line, no one would have time to think about alternative solutions. Innovation only happens if people are free to innovate, and being free to think and exper-iment is, at first glance, an inefficient use of resources. Knowing this, some industries have developed metrics to quantify the process of innovation, like the percentage of successful experiments and the overall resources invested in innovation. Being able to recognize when a project will not succeed and "failing fast," by quickly pivoting resources to another project, is an efficient way to manage the intrinsically inefficient innovation cycle.

What concerns me the most when we fail to build efficient systems is the production of unnecessary waste, especially when harmonic co-existence in the world is at

---

15. https://www.inc.com/howard-tullman/bookstores-are-back. html (Last accessed in June 2025).

stake. An analysis through the lens of efficiency enables us to dig deeper into systems that we take for granted and question their impact in our lives. The remaining chapters of this book will examine systems across several domains as objectively as possible and evaluate their efficiency and their impact in society.

Although I have my opinions about the Jack Dorsey tweet NFT, we can set judgment aside and do a technical analysis of the robustness of the NFT market. Unlike physical assets or myths with intrinsic value, such as stocks of publicly traded companies, the scarcity of NFTs is completely artificial. Their value depending solely on hype makes the market for NFTs fragile, which was confirmed empirically by the sharp drop in price for Jack Dorsey tweet and many other NFTs. As we discussed before, a fragile system is never efficient. This technical argument should inform us on how to think about the system behind NFTs. Fragile systems neither generate value nor enhance our human experience. We'll make similar analyses for every system we discuss throughout the book.

Communication is the foundation of our social structures and what enables us to create and convey myths. It is also how we interact with computers to define abstractions, such as sorting algorithms and Robotaxis. We'll start our investigations in the next chapter by analyzing the efficiency of communication systems. These will provide the building blocks for us to understand social networks.

2.

Communication

The story of the little blue house, extrasensory perception, information diffusion, and social media

"The world was so recent that many things lacked names, and in order to indicate them it was necessary to point."

Gabriel Garcia Marquez,
One Hundred Years of Solitude

I grew up hearing a joke from my grandmother, in which she would ask us "Do you want to hear the story about the little blue house? If you say yes, I'll tell you and otherwise I won't." If my sister or I answered *yes*, she would say that she "hadn't asked about *yes*," and would again repeat the same line "Do you want to hear the story about the little blue house? If you say yes I'll tell you and otherwise I won't." This would continue for a very long time, no matter what we answered. Like an infinite game, in which the only purpose is to annoy everyone who is playing, in this case my sister and I, with the same question repeating itself over and over independently of the provided answer. We'd end up begging her *"Please stop"* and she would reply "I didn't ask about *please stop*. Do you want to hear the story...."

The tradition persists in my family. My daughters now reply "Is *the little blue house IHOP?*"[1] and my mother continues "I didn't ask if *the little blue house is IHOP.* Do you want to hear the story…" My mom even promised them that she will write the real story of the little blue house in her will, so after her death the mystery may be finally revealed. A similar infinite joke was used by the Buendia family in Gabriel Garcia Marquez's book *One Hundred Years of Solitude,* during the insomnia plague. They would sit around the dinner table and ask "Do you want to hear the story about the capon?" and the conversation would last days and nights.

While it may seem irritating, the infinite joke about the capon may be an efficient way to pass time as a family when no one can sleep. In the absence of an insomnia plague, people tend to prefer more direct forms of communication. And, although we are far from the early days of the fictitious town of Macondo, when things had no name and had to be identified by pointing, communication mistakes today are still common. For example, in 1999 a spacecraft was lost due to an information mismatch between two NASA teams working on the project.[2]

---

1. IHOP stands for International House of Pancakes, an American restaurant chain that has stores that look like little blue houses.
2. https://www.simscale.com/blog/nasa-mars-climate-orbiter-metric/ (Last accessed in January 2025).

Lockheed Martin Astronautics, located in Colorado, was responsible for designing and building NASA's Mars Climate Orbiter. It used the Foot-Pound-Second (FPS) system of measurement for the spacecraft controls. Although historically FPS has been associated with England, due to the imperial system of measurement, currently England has adopted the metric system and only the US, Liberia, and Myanmar remain using FPS.

The other team working on the project was NASA's Jet Propulsion Laboratory (JPL), in California. The JPL team worked on the navigation system and used the metric system, as customary for space missions. The Colorado team expressed force in Pound-force, while the JPL team expected it was in Newtons. One Newton equals approximately 4.448 Pound-force. The orbiter cost $125 million dollars to build. It was launched in December 1998 and after 10 months, when it was about to enter Mars' orbit, the miscalculation pushed the spacecraft too close to Mars' atmosphere, causing it to burn and break into pieces.

This costly accident could have been avoided had simple validations been in place to reduce the fragility in the system. The Mars Orbiter accident was not the first nor the last caused by incompatible measurement systems. Air Canada flight 143, a passenger flight between Montreal and Edmonton, ran out of fuel on July 23, 1983[3]. Due to mismatches between the metric

---

3. https://en.wikipedia.org/wiki/Gimli_Glider (Last accessed in January 2025).

and imperial systems that produced wrong calculations, it was carrying only 45% of the required fuel for the trip.

This incident became known as the Gimli Glider, because Captain Robert "Bob" Pearson, being familiar with flying techniques rarely used in commercial flights, was able to "glide" the plane and land it on the RCAF Station Gimli, a non-operating air force base. Thankfully the accident had no fatalities. All 69 people onboard survived and only 10 had minor injuries. However, the board investigating the incident recommended that Air Canada fully convert all aircrafts to the metric system, noting that a single measurement system was more robust than managing a mixed fleet.

Today, the metric system, also known as the international system of units (SI), is widely used throughout the world. Even the US, which is one of the only three countries that doesn't fully adopt SI, requires that all government agencies use the metric system. More than 30% of products manufactured in the U.S. now use the metric system and some industries, such as pharmaceuticals, have fully switched. Having a common understanding about what things mean adds robustness, but are there any limits on how efficient communication can be? One option is to avoid communication altogether and rely on guessing, as we'll see next.

Extrasensory perception (ESP)

One task that humans are better at than computers is labeling: looking at an image and assigning labels to it. When looking at Leonardo da Vinci's Mona Lisa, some possible labels that come to mind are woman, smile, enigmatic, calm body, looking at me, and empty hands. As the common saying is that one image is worth a thousand words, there is no shortage of labels our creative minds can produce. Labels on all sorts of data are important for AI's learning algorithms. For instance, if we label movie recommendations with a thumbs up or a thumbs down, the algorithm will learn our preferences and the recommendations will be refined.

Labeling images was a particularly important task in the early 2000s, as the labels could be used to help us search for images. The search query "enigmatic woman looking at me" could retrieve an image of the Mona Lisa, given it was properly labeled with some of these terms. Since then, AI has evolved to rely less and less on human-generated labels, as we'll discuss in Chapter 6, but let's set that aside for now and let's have a flashback moment. In 2003, computer scientist Luis von Ahn came up with a creative way to label images. He invented the ESP game.

The game works as follows. There is no way users can communicate except through the game interface. When a user logs in to the game's website, they are randomly matched with another user. The pair is then presented with a single image and each user has to independently create a label for that image. They start typing words and once they've both typed the same word, even if not at the same exact time, that shared word will become the label of the image. After a successful match they move to the next image. The objective is to label 15 images in two and a half minutes, producing one new label for each.

When an image is successfully labeled, the new label becomes a taboo word. This means that it will not be accepted as a label when the same image is shown again in the future. The next time that image is used in the game users will also see its list of taboo words, and will be informed that these taboo words cannot be used as guesses. Taboo words guarantee that the same labels are not rediscovered and that a new one will be produced every time users are successful in finding a match. The game also had mechanisms that prevented cheating. It showed test images, with a predefined set of labels, to new users to check if their answers were valid. It also only stored a new label for an image after it came up as a solution a few times. This greatly reduced the likelihood of incorrect labels. If p is the probability of a round of ESP producing the wrong label, waiting for r rounds before attributing the label to the image would reduce the probability of the label being incorrect to p^r. For $p = 20\%$, with only three rounds the combined probability of wrongly producing a label would be only 0.8%.

The original paper[4] claimed that users on average produce 3.89 labels per minute playing the game. At this rate, 5,000 users playing the game simultaneously for a month could have labeled every image in Google's index back in 2004, when the paper was written. Google eventually licensed the ESP Game. Players didn't necessarily know or care that they were helping improve the quality of image search. They simply loved playing the game. Luis von Ahn created a few other games that extend ESP, including ones that ask the users to mark the region of the image that corresponds to the label, or select which of two images is the most beautiful.

In 2007 he proposed reCAPTCHA[5], which is a variation of the CAPTCHA test. Widely used across many websites, CAPTCHA distinguishes humans from robots by presenting an image with a distorted text and asking users to type it. Both CAPTCHA and reCAPTCHA are similar in concept to Turing tests. Proposed by Alan Turing, a Turing-test verifies a machine's ability to behave like a human. Turing referred to these tests as the "imitation game," hence the title of his biopic.

reCAPTCHA works like CAPTCHA, but uses fragments from old books that are being digitized. These fragments are often illegible and cannot be processed by

4. Labeling images with a computer game, Luis von Ahn and Laura Dabbish, CHI '04: Proceedings of the SIGCHI Conference on Human Factors in Computing Systems.
5. https://en.wikipedia.org/wiki/ReCAPTCHA (Last accessed in January 2025).

scanning character recognition algorithms. In the same way that ESP game users didn't necessarily know they were labeling images, reCAPTCHA users don't know they are helping digitize old books. The goal remains the same. Gather high-quality human input to improve the system. Once many people type the same character sequence for an image, there is a high likelihood that their solution corresponds to the original text. This means reCAPTCHA is antifragile – the quality gets better as more people use the system.

In both cases, users are voluntarily completing tasks that improve the system. They have fun playing ESP and help deter spam bots with reCAPTCHA. These are examples in which humans are cooperating to complete tasks, even without direct communication. Although this is a clever scheme, most tasks cannot be done through telepathy and require dialogue. And they require the exchange of messages, represented in computer bits. Even in ESP, although messages are not sent directly between players, they are exchanged indirectly through the game. Efficient communication requires precise and compact messages. Let's return to how efficient communication can be, starting with a nostalgic example from when I was a student.

Do not study

When I was in graduate school at the University of Waterloo, I often visited a donut cafe where every table had a sign that read "Do not study." My advisor, whose field of expertise was formal languages, always complained that it was sending the wrong message to students. According to him, "Do not study here" would be the correct wording. It is safe to assume that UW students interpreted the message correctly and their GPA didn't drop because of that misleading sign.

Attributing the incorrect meaning to a message could, however, have unintended consequences, such as losing a spacecraft or a plane not having enough fuel. It is difficult to know how another person interprets a message, especially when they lack context. Having a common context allows people to communicate more effectively. For example, unifying engineering projects under the metric system can remove ambiguities in communications. Organizations with clear values and expectations enable employees to collaborate more effectively. However, despite our best efforts, some level of communication dissonance will always exist. And although we cannot eliminate it completely, a shared context and precise communication helps to minimize it. Precision in communication helps organizations achieve more and avoid rework, a topic we'll explore further in Chapter 5, when discussing organizational efficiency.

One example of unambiguous communication is using programming languages to describe algorithms. Programming languages have a formal structure that allows computers to perform the steps of the algorithm in precise order. Figure 2 shows the bubble sort algorithm, which we discussed in Chapter 1, written in the

C# programming language. There is no need to understand the details of this C# program. Figure 2 is just for illustration, and it's not required for understanding the rest of the book. The important point is that the description is precise and any computer that executes this implementation of bubble sort using the same list as the input will produce the same result as the output. Similar to the concept of replication in science experiments, this consistent repeatability is known as **determinism**. We often strive for determinism in computer systems because it increases robustness. Predictable behavior makes it easier to detect, reproduce, and fix errors. Not all computer programs need to be deterministic to be correct and useful though. For example, a random number generator or a function that returns an arbitrary permutation of a list shouldn't always produce the same output, because that would defeat the purpose.

With AI, we are moving away from programming languages and closer to spoken, natural languages. Today, you can ask an AI program in English to sort a list of numbers and get a correct result. While programming languages aren't going away, more and more tasks will be expressed to computers in natural language over time. AI will enable us to move away from formalism in human-computer communications and move toward a world in which computers operate more like humans.

```
1      public int[] SortList()
2      {
3          var n = list.Length;
4
5          for (int i = 0; i < n - 1; i++)
6             for (int j = 0; j < n - i - 1; j++)
7                if (list[j] > list[j + 1])
8                {
9                    var tempVar = list[j];
10                   list[j] = list[j + 1];
11                   list[j + 1] = tempVar;
12               }
13
14     return list;
15     }
```

Figure 2. Bubble sort algorithm described in C#. The variable *j* identifies the location in the list, also referred to as the index, of the element we are comparing with its adjacent neighbor in index *j +1*. The value of the element in index *j* in the list is *list[j]*. Similarly, *list[j + 1]* is the value for index *j + 1*. The fragment in lines 9 to 11 swap the list elements using variable tempVar as a temporary buffer. We store the old value of *list[j]* in this temporary variable. Then, we assign the value of *list[j+1]* to *list[j]*. If we hadn't saved *list[j]'s* value in the temporary variable, it would have been lost after line 10. Finally, in line 11 we assign the old value of *list[j]* to *list[j + 1]*, effectively swapping the values. The *"for"* instructions in lines 5 and 6 control how many times we do the comparisons and swaps. We affectionately call these instructions for loops.

In many ways, AIs are more like superhumans. AI systems have read and memorized a lot more text than any of us ever could, and, amongst other things, can write impressive essays about almost any topic and even draw very realistic pictures. In many other ways, they are still like babies. They lack common sense, need constant supervision, and can easily misinterpret you. One of the main hurdles to making AI more well-rounded is that AI requires context. Although it is very easy for the UW students to understand that "Do not study" means "Do not study here," this is much harder for computers.

The difficulty is simply due to their lack of common sense. Although AI systems have already "learned" all of Wikipedia and every piece of publicly available information out there, they don't know that if you are in a university donut cafe with a limited number of tables, you should not study there when the place is busy and other customers want to eat donuts and drink lattes. Understanding context and common sense remains a major hurdle in AI, one that top researchers, like Professor Yejin Choi of Stanford, are actively exploring.

Prof. Choi, who is a recipient of the MacArthur Foundation "Genius Grant," recently explained what common sense means for AI in a New York Times interview:[6] "It's the unspoken, implicit knowledge that you and I have. It's so obvious that we often don't talk about

---

6. https://www.nytimes.com/interactive/2022/12/26/magazine/yejin-choi-interview.html (Last accessed in January 2025).

it. You and I know birds can fly, and we know penguins generally cannot. So AI researchers thought: we can code this up. Birds usually fly, except for penguins. But in fact, newborn baby birds cannot fly, birds covered in oil cannot fly. The point being, exceptions are not exceptional, and you and I can think of them even though nobody told us. It's not so easy for AI."

With or without common sense, computers play a central role in how we communicate today. When we talk about communication we need to consider not only human-to-human interactions, but also human-to-computer and computer-to-computer interactions. Like in the children's telephone game, in which a message is whispered from one person to the next in a circle, the larger a message is, the harder it is to both convey and interpret. In telephone, a very large message will almost certainly get distorted. The same is true even if the message is short but the number of people playing telephone is large. While computers are much better than humans in processing large messages, efficiency remains paramount even for computer-to-computer interactions. Especially when thousands of computers need to exchange a large number of messages to execute a task.

Playing telephone with a computer

As we discussed in Chapter 1, computers only understand sequences of zeros and ones. Playing telephone with bit messages would for sure fail with human participants, as no one would be able to properly deliver a long and unmemorable message like 0001101011100. Computers, however, fare much better at bit telephone than humans do. Let's start by understanding bits in more detail. A single bit can represent two values, 0 or 1. With two bits, we can represent four values: 00 is zero, 01 is one, 10 is two, and 11 is three. If we continue on like this, we'll see that with three bits we can represent eight values, and with n bits we would be able to represent 2^n values.

Since the English alphabet has 26 letters, we need at least five bits to fully represent all letters. With only four bits, we could only represent a maximum of 2^4 or 16 letters. Five bits gives us 32 values, so we can even cover six extra elements, such as space, punctuation, and end of line. In reality, the standard representation for letters uses one byte,[7] which is eight bits, so we can also represent capitalized letters, numbers, and special characters, such as "+" and "&".

According to Wikipedia,[8] the longest word in English is the chemical composition of titin, the largest known protein.[9] It has 189,819 letters and attempts

---

7. The standard representation for letters is ASCII, which is an acronym for American Standard Code for Information Interchange.

8. https://en.wikipedia.org/wiki/Longest_word_in_English (Last accessed in January 2025).

9. https://en.wikipedia.org/wiki/Titin (Last accessed in January 2025).

to say the entire thing have taken from two to three and a half hours. It'd be a very boring telephone game. Considering the more compact but unorthodox representation of 5 bits per letter, the titin protein stated in a message would have 189,819 × 5, which is almost one million bits.

It should be clear that if we want the telephone game to succeed, when a player receives a message of n bits as the input, it should send all the n bits to the next player. Missing only a single bit would lead to the wrong message being delivered. However, if we want to encode text messages, we can be smarter about how to interpret bit sequences to achieve smaller messages. One possible technique is to encode text using the frequency of letters in written language as a guide. E is the most popular letter in English, appearing in about 12.7% of the texts, followed by T (9.1%), A (8.2%), and O (7.5%). The least popular letters are Q and Z, with about 0.1% of representation each.

Instead of using five bits to represent each letter as I proposed before, we can improve our bit representation of titin by incorporating letter frequencies. The insight is to use a variable size scheme, with fewer bits to represent the most popular letters, such as letter E, at the expense of using more bits for unpopular letters such as Q and Z. This technique, known as Huffman coding,[10] is a

---

10. https://en.wikipedia.org/wiki/Huffman_coding (Last accessed in January 2025).

popular compression algorithm. Many of the widely used computer formats, like JPEG and MP3, use variations of Huffman coding.

The main trick in assigning letters to bit sequences in Huffman coding is to avoid ambiguities. For instance, assigning 0 for E, 1 for T, and 10 for A would not work, since we would not know how to decode the message 0101. It could be decoded as two different messages: ETET or EAT. To avoid this problem, the Huffman algorithm assigns bit sequences with unique prefixes. E becomes 00, T 01, A 100, and so on. With this encoding, we would be able to unambiguously decode 0101 as TT. In this representation with unique prefixes, the most frequent letters, E and T would be represented in only two bits, while Z would take eight bits.[11]

We can extend the frequency of letters trick to further improve the efficiency of our solution. If we knew that some letter sequences such as TH or HE were very popular in the original text, we could assign small bit sequences to represent these combinations. Considering only single letter frequencies, however, we can prove that Huffman coding is optimal, meaning that there is no possible way to achieve higher compression. It is also very effective. If we use it to compress Shakespeare's Romeo and Juliet, which has about 145 thousand characters including spaces, punctuation, and

---

11. Please see a visualization of Huffman coding at https://digitalagencybook.org/visualizations/huffman-coding.

new lines, we can reduce the play to about 60% of its original size.

Now, let's revisit the titin protein. Even though there are only 20 amino acids, the word for titin contains tens of thousands of them in sequence, starting with methionyl, followed by threonyl, and so on:

methionylthreonylthreonylglutaminylalanyl...

There are in total 34,350 amino acids in the sequence. One way to compress the word is to use amino acid abbreviations, such as "M" for methionine and "Q" for glutamine, instead of the whole word. That would reduce the number of characters from 189,819 down to 34,350. A compression of 5.5 times. That compression is much more substantial than what we got from Huffman coding on Romeo and Juliet. Our domain knowledge on amino acids led us to this better compression method, since we know that the titin protein is a sequence of only 20 possible amino acid types. Huffman coding, on the other hand, does not benefit from any domain knowledge and simply relies on frequencies that are computed from the data. It knows nothing about what the letters or words mean.

If we look at the frequency distribution of the amino acids in the titin protein, however, it looks more uniform than the distribution of letters in English texts. Figure 3 shows both distributions.

The most frequent amino acid in titin is glutamate, which coincidentally is represented by the popular letter E. The least frequent is methionine. Glutamic acid (E) appears 8 times more often than methionine (M). In English texts, E appears 181 more times than Z. This means that if we now apply Huffman coding to the word composed of the amino acid letters, we'd achieve fewer gains than when we apply it to English texts. It would be about 18% gain, as opposed to 40% in the case of Romeo and Juliet. This 18% is calculated by representing each of the 20 amino acid types using a Huffman code and using it to compress the titin, similar to what we did to Romeo and Juliet. The more skewed the distribution of letters is, the better Huffman coding performs. In the uniform case, where all frequencies are similar, its benefits are greatly reduced.

Frequency (%) vs. amino acid in titin

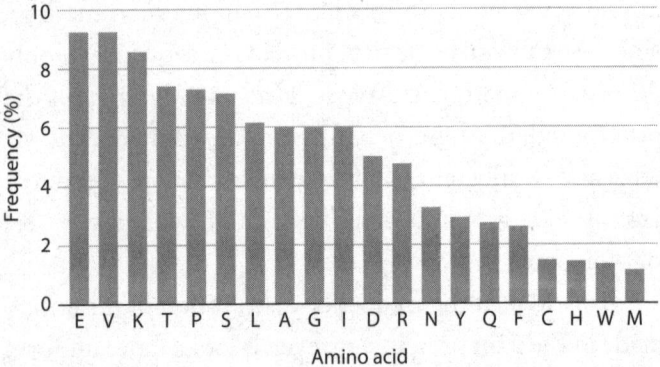

Frequency (%) vs. letter in English texts

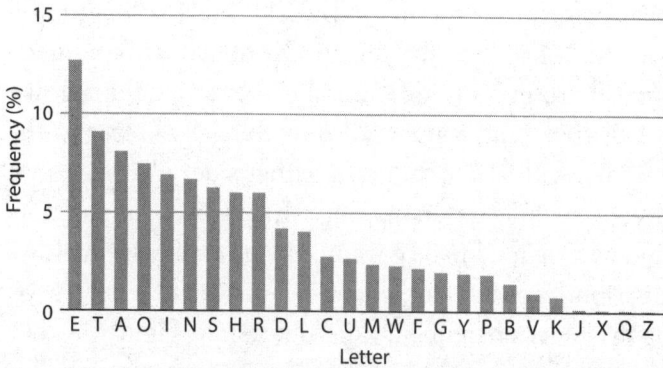

Figure 3. Distribution of amino acids in titin proteins and letters in English texts.

This example highlights that by understanding the domain, such as protein structures in this case, and designing the correct abstractions, we can be a lot more efficient. By understanding that the largest English word is really just a combination of only 20 amino acids, we can achieve a compression of 5.5 times. This can then be improved by an additional 18% with Huffman coding. The end result is a sequence of about 141 thousand bits for our telephone game, instead of the original uncompressed sequence of almost one million bits. This is about 6.7 times more efficient.

Another subtle point lies in the comparison of the two distributions shown in Figure 3. The natural world, governed by chemistry, physics, and biology, tends to operate in normal distributions. This is not only true for the amino acids in titin, but for most of the distributions that come from the natural world, such as height and weight distribution for humans. Normal distributions are spread around a middle number. They are shaped like a bell curve and show that most things are close to the middle, with a few things on either side. The myths we create, such as letters in English texts, city population, and wealth, tend to be governed by power laws, which are distributions that have a long tail instead of a bell curve, as in the letter frequency distribution in Figure 3. They tend to be much harder to predict and rationalize. This is one of the main reasons that computer systems and financial markets are more prone to black swan events than physical systems: it is much harder to reason about power laws than normal distributions.

Our telephone game showed us how text can be represented in computer systems and what types of efficiency gains are possible. Memory athletes draw from the same type of cleverness used to conceive Huffman coding, and other algorithms, to memorize large sequences of numbers.[12] Both humans and computers need these techniques to represent information efficiently. With that understanding, we can now move on to investigate how information propagates in a connected network and in society.

---

12. https://en.wikipedia.org/wiki/Memory_sport (Last accessed in May 2025).

Information cascades

When playing telephone, every player tries to correctly convey the message they heard to the following player, unless they are purposely trying to sabotage the game. When collaborating in teams or when interacting in social media we are not simply repeating information. We are adding to the conversation. This makes the problem a lot more interesting, as each person's choice may influence what others do. In many situations, individual behaviours are aggregated to produce collective outcomes. Professor Easkey, an economist, and Professor Kelinberg, a computer scientist, both from Cornell, wrote the authoritative book on this topic.[13] Most of the ideas presented in this section are described more formally there.

Let's start defining the problem of information cascades. Consider the simple case of a tourist that just arrived in town and wants to have breakfast. She wants to go to the Green diner based on the recommendations she got from friends. However, she realizes that while the Green diner is completely empty, the Orange diner across the street is packed. Not only that, there are people waiting in line. The fact that others have chosen Orange over Green may impact her decision to eat at the Green diner and cause her to ignore her friends' recommendation. If she switches her mind, she will do so because she

---

13. *Networks, Crowds, and Markets: Reasoning About a Highly Connected World*, D. Easley and J. Kleinberg, Cambridge University Press, 2010.

believes the information she learned from Orange diner patrons, in aggregate, is stronger than the information she obtained on her own. This is herding in action.[14]

Cascades can be understood by witnessing the social phenomenon of herding. Herding may happen in any situation in which people can observe the other people's past or current actions. In real life, this "herd mentality" includes acts such as liking a social media post, buying stocks, or standing in line at a diner. In information cascades, we are not talking about social pressures that may influence people to change their minds, although that possibility is always present and likely a factor to be considered as well. Even if the tourist is completely rational and doesn't feel any social pressure, she may genuinely believe that all the folks standing in line at the Orange diner may collectively know more than her about how the Green and Orange diners differ in quality.

But exactly how many Orange diner patrons are required to stand in line to make the tourist ignore her own research and follow the crowd? Let's try to understand this in a bit more detail with another example.[13] There is an urn in front of the classroom with three marbles. The teacher declares that there is a 50% chance that two of the marbles are red and one is blue and 50% chance that two of the marbles are blue and one is red. The rules of the game is that students will form a line,

---

14. A Simple Model of Herd Behavior, A. Banerjee, The Quarterly Journal of Economics, Vol. 107, No. 3, 1992.

and when it's their turn, each student will draw a marble from the urn, look at it, and place it back in the urn without showing it to the other students.

After placing it back into the urn, the student will choose between two options, "majority red" or "majority blue," and publicly announce their choice to the class. At the end of the game, students that guessed correctly will receive a monetary reward, while they will get no reward for guessing wrong. Let's assume the students are rational and want to receive the reward. They also have no incentive to cheat. Although this example is fairly unrealistic, it mimics the scenario of the Green and Orange diners in several relevant aspects.

First, each student has information that only they know. The color of the marble they drew. This is equivalent to the research the tourist did by asking for recommendations of diners from her friends. Second, each student publicly announces their choice. In the diners scenario, they also publicly announce their choice by standing in line or by simply eating at the diner. Third, there is order in the game. All the students in line can learn the choices of the students that went before them and already declared their guesses. Similarly, every new patron can look at how many customers are in the Orange and Green diners before choosing where to have breakfast. Finally, there is a decision to be made based on the information each student gathered so far, the color of the marble they drew and the choices from other students. The tourist also has to make a decision based on her own research and on her observation about the behavior of other hungry patrons.

Let's analyze what happens after each student makes their choice:

- The first student, after drawing a marble, should make the rational choice of guessing the color of the marble they drew. If there are two red and one blue marble, the probability of drawing red is twice that of drawing blue. Guessing red if their draw was red would be the intuitive and rational choice. Let's assume the first student chooses red.

- If the second student also draws a red marble, their decision is simple. Otherwise, they have a harder choice. They can safely assume that the first draw was red and they know for a fact that their draw was blue. The second student then may correctly assume that both options have equal probability and just make a wild guess. However, although the second student won't personally benefit from this decision, if they want to convey more information to the rest of the students, they should guess their draw and pick blue. Otherwise, if they guess red, the rest of the class may incorrectly assume the second draw was also red.

- We'll call a guess a "true guess" when the student uses their private information, the color of their draw, to make their choice. When we get to the third student, we can assume the first two students made true guesses, each announcing the color they drew. If the first two guesses were red, the third student should also guess red, independently of their draw. In this scenario the third student correctly assumes that with observations either red-red-red or red-red-blue, "majority red" is the most likely outcome. The fourth student will follow the same line of thought, and also guess red. The same will be true for every subsequent student. Notice that only two consecutive draws of the same color were enough to ignite the cascade.

- If the third student hears two different guesses, red-blue, they are again in the same situation as the first student. To maximize their chances of winning the reward and to convey the maximum amount of information to other students, they should make a true guess and declare whatever color they draw.

- The fourth student that hears red-blue-red or red-blue-blue knows that all the guesses so far have been true guesses. They are in a similar situation to the second student. They should also make a true guess. If they draw the same color as the previous student, they will start a cascade.

- At this point we have either started a cascade or have seen pairs of alternating true guesses red-blue-red-blue or red-blue-blue-red. This alternating pattern will continue until two marbles of the same color are drawn. When that happens the cascade will start.

Although this is not very intuitive, we have seen that only two consecutive draws of marbles of the same color are enough to start the cascade, after we discount pairs of alternating colors. If we compute the probabilities, there is a 55% chance that the cascade will start on the third student. And after that, there is a 55% for the cascade to start at each subsequent step, given it hasn't yet started. This means that in just a few steps the probability of starting a cascade will asymptotically increase to 100%. Moreover, once a cascade starts, there is a 20% that it would lead to the wrong outcome, guessing red instead of blue or vice-versa.

The way I describe the game assumes that all students are rational beings. They are maximizing both their chances of winning the reward and the amount of information they convey to the rest of the students in class. The field of behavioral economics debunked the notion of the homo economicus, meaning rational humans, by demonstrating that humans behave irrationally in many situations. Daniel Kahneman's *Thinking Fast and Slow*[15] describes several experiments that demonstrate this, such as the planning fallacy, which is our tendency to overestimate benefits and underestimate costs, possibly leading to some regrettable kitchen remodels. Even though we discounted irrationality and other factors, such as peer pressure, most of the learnings from our analysis would still hold true otherwise. From this simple example we can draw some important conclusions about cascades:

---

15. *Thinking Fast and Slow*, Daniel Kahneman, Farrar, Straus and Giroux, 2011.

1. They are easy to start. As we have seen, all we need is two consecutive draws of the same color. Even for diners, few patrons waiting in line is likely enough to swing future customers.

2. They use very little information. Once the cascade starts, people largely ignore their own information, such as the color of their own draw or the diner reviews they personally collected. Only the information collected before the cascade started is used. The aggregate information for all the other participants, which may be collectively very rich, is largely ignored.

3. They may lead to the wrong outcome. In the classroom example, once the cascade starts, there is a 20% chance that it will lead to the wrong outcome. This is a direct consequence of the point above. This observation leads us to think of cascades as **fragile systems**.

   If instead of announcing their guesses sequentially, the first ten students did so at the same time, we would allow the other students to learn from the first ten observations before making a guess. This would reduce the probability of a wrong outcome to less than 8%. With twenty students it would reduce to less than 4% and with one hundred students it would be approximately 0.02%. These numbers also assume the students are acting rationally. Contrary to the original sequential scenario, this setup is **antifragile.** Increasing the number of students in the initial group reduces the chances of making a mistake.

4. They are easy to dismantle. Once a cascade has started, we can assume that two students drew the same color marbles and made true guesses. Suppose these draws were red, and that after that, all the information gathered by subsequent draws was ignored. Let's assume the teacher changed the rules after the cascade started, allowing two students to show their draws to the entire class. If coincidently these two draws were blue, the cascade would have been dismantled. Again, this is assuming the students are acting rationally.

An important factor of why cascades are so easy to dismantle is the fact that they use little information. Intuitively, people know that most of the individual knowledge each person collected, which in this case is just the color of their own marble, is not being used. This makes cascades **nondeterministic**. Small perturbations to the system may lead the system to produce different results.

Once we understand that information cascades are fragile and nondeterministic, we should expect that any system that is susceptible to herding would also have these properties. Keeping that in mind, let's now look at ways that information is commonly disseminated: books, web pages, and social media.

My only real hope was that my mother would like it

Malcolm Gladwell wrote his first book, *The Tipping Point*[16], in 2000. He was already a well-respected writer for The New Yorker, and The Washington Post before that. Many of his published articles were related to the topics covered in the book, helping to solidify his credibility as an insightful author. But despite all his success and credibility, he was uncertain about the prospect of the book's success. As he put it, "When I wrote it, my only real hope was that my mother would like it! I've considered all my books to be very private, idiosyncratic projects designed to make me happy. And I'm forever surprised when they make other people happy too."[17]

The Tipping Point likely became a success due to some of the principles described in the book itself: **the law of the few**, **the stickiness factor,** and **the power of context**. The law of the few states that a few people with rare skills are enough to start an epidemic. In the book's case, some early readers adopted the ideas and disseminated them widely. They included academics and business leaders, who also helped transform the book's title into a buzzword. That enabled the book to gain popularity through word-of-mouth. The stickiness

16. *The Tipping Point*, Malcolm Gladwell, Little Brown, 2000.
17. https://www.fastcompany.com/1800273/malcolm-gladwell-has-no-idea-why-tipping-point-was-hit (Last accessed in January 2025).

factor principle says memorable content increases retention. The book appealed to multiple demographics, presented innovative ideas, and was engaging. Finally, the power of context states that epidemics need the right cultural environment to flourish. Published in 2000, *The Tipping Point* was well timed with the rise of the internet.

But Gladwell was probably right to be initially skeptical about the success of his first book. The number of books published in the U.S. has been increasing steadily year over year. According to 2023 statistics,[18] there are between 500 thousand to one million new books published each year, excluding self-published books. This number is growing rapidly. In 2013, the number in the U.S. was 275,000, and if you include self-published books, the number gets close to four million.[19] With that many books, one must wonder how any book becomes a bestseller.

Although one million books per year seems like a ridiculously large number, it is a far cry from the trillions of web pages on the internet and the more than one billion posts made daily on social media. The reason for this gap is that there is a **high barrier of entry** in the book market. Writing a book is hard, demands time, and there are gatekeepers in the publishing industry. Publishers may

18. https://www.tonerbuzz.com/blog/how-many-books-are-published-each-year/ (Last accessed in January 2025).
19. https://en.wikipedia.org/wiki/Books_published_per_country_per_year (Last accessed in January 2025).

reject unsolicited manuscripts and may refuse even the solicited ones. Although some of this is alleviated with AI and self-publishing, writing a book requires skill and is time consuming. There are several domains in which the high barrier of entry, like the book market, provides some benefits.

The National Football League (NFL) has only 32 teams. This allows experts to get detailed statistics for every player. For example, at a given time during football season, there may be only 32 starting quarterbacks to compare and dissect, with vast amounts of information about each one. The barrier of entry is so high to be an NFL quarterback that it's likely that many talented athletes give up and don't even try, or don't advance to the requisite playing level, before the time players get drafted. However, we don't have the opposite problem, where an unskilled quarterback may be drafted based on little information. The high barrier of entry brings **determinism**. With a few players to analyze, and having quality data at their disposal, experts will likely converge into a good decision.

The NFL draft, officially known as the Annual Player Selection Meeting, is how football teams recruit players for the upcoming year. Teams prepare for the draft very carefully. As the number of potential players joining the NFL is not large, teams have all the data they need to do a thorough analysis of every candidate. High quality data about each player leads to determinism. It is unlikely outcomes for a given team would vary widely when the same players are considered.[20] The same is true for many other areas and industries, such as car models. There is a lot of information about every model of SUV you can buy and even the most detail-oriented buyers probably have access to more data that they can possibly consume.

Even though readers have many books to choose from, we have some empirical evidence that, like the NFL, the book market still benefits from its high barrier of entry. Although it is very possible that some excellent books never make it to the best seller list, the ones that make it are generally high quality and appeal to a large audience. They are unlikely to be the result of some information cascade that swayed people to read an uninspiring book. Despite the high barrier of entry, it is still possible for a debut author, even an unsuspecting one, to hit the mark. Delia Owens was a zoologist and conservationist before she wrote her first novel *Where the Crawdads Sing.*[21]

The book clearly has the stickiness factor, resonating with many readers, who were likely intrigued by the life of its protagonist, Kya Clark. It also benefited from the law of the few. The book was selected for Reese Witherspoon's

---

20. Of course, no system is perfect and there are blunders, even for NFL quarterbacks, as in the case of the 2nd pick, Ryan Leaf, in the 1998 draft. He was the second pick but his NFL career was shortened due to behavioral and other non-football related issues: https://en.wikipedia.org/wiki/Ryan_Leaf (Last accessed in May 2025).
21. *Where the Crawdads Sing,* Delia Owens, G. P. Putnam's Sons, 2018.

book club and Barnes & Noble's Best Books of 2018. That helped start the epidemic. In 2019 the book sold more copies than any other adult title, fiction or nonfiction and in 2022 a movie based on the book was released. Having sold 18 million copies so far, *Where the Crawdads Sing*, together with Michele Obama's *Becoming*,[22] are the only two books published in 2018 or after in the list of best-selling books of all time.[23]

Despite being a very memorable book, it is likely that it would have gone unnoticed if instead of one million books per year, the book market produced one thousand times more, one billion books per year. At least there is a much smaller chance Reese Witherspoon would have read it. By the same token, the NFL draft would not work with 32 thousand teams. In both cases, the high barrier of entry plays a central role in avoiding the problem of too little information we've observed with information cascades. Reese Witherspoon, book critics, and NFL coaches have plenty of options to choose from, with rich information about each, instead of an overwhelmingly large number of obscure options.

But what about book reviews? Are book reviews susceptible to information cascades? It is definitely true that before writing a review you can read all the previously submitted reviews. These reviews may influence, and even

---

22. *Becoming*, Michele Obama, Crown, 2018.
23. https://en.wikipedia.org/wiki/List_of_best-selling_books (Last accessed in January 2025).

sway, the reviewer to change what they intended to write. However, we must consider that reviews are more useful for readers that have not yet bought the book. After reading the book, the reader will acquire the information required to write an informed review. Given that, book reviews don't seem as susceptible to cascades. Reading a book is typically an investment of several hours. It is unlikely that someone will ignore their own impressions and instead go with the option of other reviewers, especially after making such an investment of time and intellectual energy.

The fact that *The Tipping Point* and *Where the Crawdads Sing* are both very successful is empirical evidence that, despite the growing number of titles being published each year, the high barrier of entry enables determinism in the book market. Well-written and impactful books will get the right attention and become successful. As book reviews are not too susceptible to cascades, it is hard for undeserving books to gain popularity. All seem fair and square. But what happens when we lower the barrier of entry and democratize publishing even more? Let's investigate the scenario of the web as an information dissemination platform.

Bringing order to the web

Since the creation of the web over 30 years ago, it has evolved into a key information system. Nowadays it is imperative that every newspaper and magazine have an online presence, with many of them even existing only in a digital format. But with over 100 trillion web pages,[24] and the rise of fake news and artificially created or enhanced content, it is hard for readers to know if a web publication adheres to journalistic ethics and standards. Journalists must be as accurate as possible when conveying information by including correct and accurate information while not excluding important information, verifying sources, writing in an easily understandable manner, and so on. While we can expect some level of responsibility from legitimate publications, we cannot guarantee the integrity or accuracy of content found on blogs and web pages in general.

One of the key differentiating features of the web is its use of hyperlinks. In the same way a research publication cites relevant studies, hyperlinks also provide some credibility to the web page being referenced. In academia, the number of times a published work is cited is a quantitative measure of its significance, and there are metrics in place in academia to evaluate researchers that contemplate the number of citations of their publications.

The h-index is one of such metrics that is widely used for evaluating credibility, being even used for tenure considerations. A high h-index means the researcher has a high number of well cited papers, while a low h-index indicates that their papers are not frequently cited, or they have few published papers, or both. A researcher has an

---

24. https://searchengineland.com/googles-search-index-es-hits-130-trillion-pages-documents-263378 (Last accessed in January 2025).

h-index of 10 if they have at least 10 papers with 10 or more citations. If they have a widely cited paper, such as 2000 citations, but if all of their other papers have only 10 citations, their h-index will be still 10. The algorithm to compute the h-index is the following:

1. Sort the publications by descending order of citations.
2. Find the highest number h for which h papers have at least h citations.

If we were to write the algorithm to compute the h-index in computer code, we would have to rely on a sorting function, hopefully quicksort, to sort the publication list by citation in descending order. We may arrive at a sorted list of citation counts like 202, 148, 97, 52, 7, 5, and 2 in the case of a junior researcher. We would then go through this list, and for each element we inspect, we'd check if the number of citations is higher than the number of papers considered up to that point.[25] In this case:

Step 1. 202 is greater than 1, then h-index = 1.
Step 2. 148 is greater than 2, then h-index = 2.
Step 3. 97 is greater than 3, then h-index = 3.
Step 4. 52 is greater than 4, then h-index = 4.
Step 5. 7 is greater than 5, then h-index = 5.
Step 6. 5 is not greater than 6. Stop and return h-index = 5.

---

25. Please see a visualization of the h-index algorithm at https://digitalagencybook.org/visualizations/h-index.

In the case of the web, we can follow a similar model and use citations as a validation of credibility. Intuitively, if a web page describing how to cook a Thanksgiving dinner has many incoming hyperlinks, there is a high likelihood that it is credible and it actually provides good recipes because it signals trust and value to other users. If the links come from other credible sources it is even better. This is the idea behind the **PageRank** algorithm,[26] proposed by Google cofounders Sergey Brin and Larry Page while they were graduate students at Stanford. The algorithm name refers to both Larry's last name and the term "web page."

The logic behind PageRank is that, starting from a random web page on the internet, we can compute the probability of reaching any other page by clicking on hyperlinks. The probability of reaching page W is W's PageRank. PageRank for W can be determined by two factors: the number of links to W from other web pages and the PageRank of those web pages that link to page W. Pages that have a higher number of incoming links would be assigned higher probabilities, as we'd have more ways to reach it by clicking on hyperlinks. Additionally, a page that is linked by a source page that itself has a high PageRank, will benefit from that link and inherit part of the source's page PageRank. If there is

26. The anatomy of a large-scale hypertextual Web search engine, S. Brin and L. Page, Computer Networks and ISDN Systems. 30 (I1–7), 1998.

a link from John Doe's blog, which has a low PageRank, to web page W, it will benefit from the link but only marginally. A link from The New York Times, which has a high PageRank, would contribute much more assertively to page W's PageRank.

There are many variations of the PageRank algorithm, but the principles described here are common to all flavors of the algorithm. PageRank is computed in several steps, each of which transfers some value of PageRank amongst the pages. In the description below I use *new* and *old* to represent the PageRank values of different steps of the algorithm. After each step, the *new* values become *old*, and we recompute the PageRanks with the refined numbers. The sum of all PageRanks derived in every step must always add to one, as the values represent the probability of arriving at each page. They start with equal values, which are gradually refined.[27]

Let's consider an example with four pages, A, B, C, and D. A has links to B, C, and D. B has links to C and D. C links to D. D links to A, as shown in Figure 4. Multiple links to the same page are counted only once and links from a page to itself are ignored. Initially all the PageRank values are set to 0.25, as we assume an equal chance of reaching each page to begin with. The PageRank for D can then be computed by:

---

27. Please see a visualization of the PageRank algorithm at https://digitalagencybook.org/visualizations/pagerank.

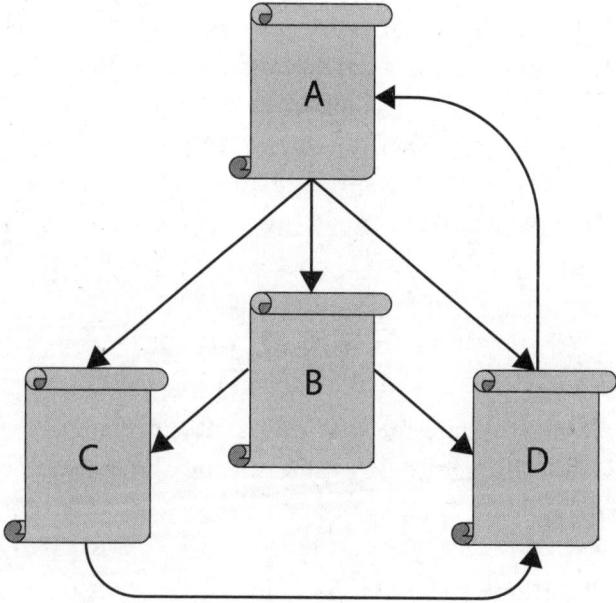

Figure 4. Example of four pages, *A, B, C,* and *D,* with hyperlinks amongst them.

$$\text{PageRank}_{new}(D) = \text{PageRank}_{old}(A) / 3 +$$
$$\text{PageRank}_{old}(B) / 2 + \text{PageRank}_{old}(C)$$

where the PageRank of each page is divided proportionally to all the target pages it has hyperlinks pointing to, as the intuition behind PageRank is that it is the probability of reaching a given page by following links. With n links, we have probability $\frac{1}{n}$ to follow each link. Since

A has links to the three other pages, its PageRank value is divided by three, while *B*'s contributions are divided by two. A similar formula can be written for the other pages:

$$PageRank_{new}(A) = PageRank_{old}(D)$$

$$PageRank_{new}(B) = PageRank_{old}(A) / 3$$

$$PageRank_{new}(C) = PageRank_{old}(A) / 3 + PageRank_{old}(B) / 2$$

If we compute the new values after one step, using 0.25 for all $PageRank_{old}$ values in the above formulas, we'll get:

$$PageRank_{new}(A) = 0.25$$

$$PageRank_{new}(B) = 0.0833$$

$$PageRank_{new}(C) = 0.2083$$

$$PageRank_{new}(D) = 0.4583$$

By iterating over these formulas for multiple steps, each time using the updated values computed in the previous step, the page ranks will converge to stable values. This means that, eventually, performing extra steps won't change the values in a meaningful way. In this simple example, about 10 steps are enough, leaving us with:

$$PageRank_{final}(A) = 0.3498$$

$$PageRank_{final}(B) = 0.1175$$

$$PageRank_{final}(C) = 0.1777$$

$$PageRank_{final}(D) = 0.3550$$

Web page D has the highest PageRank as it has links from all other pages. Page A has the second highest PageRank as it has an incoming link from D. Pages B and C have lower PageRank values as they have fewer incoming links. As described by Brin and Page in their original paper, PageRank brings order to the vast number of pages on the web. In this example the order would be D, A, B, C. PageRank greatly improved the quality of web search results by influencing the order search results are displayed, and was a key differentiator of Google over the search engines that preceded it. PageRank and its many variations help us identify web pages that are authoritative and remove at least some of the manual labor and diligence required by search users to determine if the returned pages are reliable. The list of which sites Google considers to have the highest PageRank values is not publicly available. All search engines have their own lists based on which variation of the algorithm they use. However, it is very likely that despite the variations, the top five PageRank websites are: google.com, youtube.com, facebook.com, wikipedia.org, and amazon.com.

This is estimated due to the number of links pointing to these sites from authoritative sources.

Let's return to John Doe's blog and the New York Times. Which one should Google rank higher in your search results? Algorithms like PageRank help promote the most authoritative source. The New York Times, in this case, is the one who likely has the highest number of incoming links from other pages with high PageRank. Search engines use the PageRank value as one of the sorting criteria when returning results to a search query. This leads us to believe that The New York Times content would be prioritized over the blog. Another important aspect of PageRank is that it is not subject to information cascades. That would happen if web page authors, by observing that one given web page W is becoming popular with an influx of incoming links, would be swayed to also add new links to W.

Unless in the case of fraud, in which a website owner may pay for fake links, or devise other schemes to boost its PageRank, maximizing its chance of being shown as a result in search engines like Google and Bing, the PageRank algorithm is very deterministic. Despite new sites and web pages being created every day, we already have a very large number of links from existing pages to analyze and it is unlikely that the daily changes will be significant enough to drastically alter PageRank values. And although fraud is always a factor, search engines have entire teams of engineers dedicated to detect, understand, and mitigate fraud.

This gives us some reassurance that, even with more than one trillion web pages available, there are reliable mechanisms search engines use to distill authoritative sources of information. In the same way researchers cannot easily fabricate citations to their work, website owners cannot easily make their pages more authoritative. Social networks, however, have substantially changed how we consume information. Let's start by understanding how information spreads in a connected community.

The strength of weak ties

In 1973, Stanford Professor Mark Granovetter published his groundbreaking paper, *The Strength of Weak Ties*.[28] It defined weak ties as acquaintances, colleagues, and distant friends, and strong ties as family and close friends. Granovetter showed that weak ties play a crucial role in connecting different groups. They are the connecting bridges that enable ideas to spread. The paper has been very influential in several areas, including economics, political science, business, and technology, receiving more than 70 thousand citations over the years. Interestingly, 90% of those citations came after 2000, with the rise of the internet and social networks and their impact on how people interact.

In networks with only strong ties, people already share information amongst themselves. Think about

---

28. The Strength of Weak Ties, Mark Granovetter, American Journal of Sociology, 78(6), 1973.

an isolated family or community that does not interact with anyone from outside. It would be unlikely that members of that community would diverge in the way they think. Everyone eventually would have the same information, with no new ideas entering or leaving the community. Weak ties enable access to new types of information. In his seminal paper, Granovetter showed that weak ties play a crucial role in many areas, including job searches.

The paper presented a study on 282 job applicants in the U.S., and found that acquaintances were more helpful in job searches than the applicants' close connections. In 2022, a group of researchers from LinkedIn, Stanford, MIT, and Harvard published a study in the prestigious journal Science[29] that also confirmed the strength of the weak ties theory for job searches using extensive data from LinkedIn over a five year period. During this period, 2 billion new ties and 600 thousand new jobs were created. The researchers controlled the prevalence of weak ties for over 20 million people through LinkedIn's "People You May Know" feature, which suggests new connections to LinkedIn users based on common connections or followings. Their data demonstrated that weak ties were more effective than strong ties in enabling job mobility, especially in more digital industries.

---

29. A causal test of the strength of weak ties, K. Rajkumar, G. Saint-Jacques, I. Bojinov, E. Brynjolfsson, and S. Aral, Science, September 2022.

Most social media platforms encourage the creation of weak ties. Although it is hard to get precise data on this, it is estimated that on average, 60-80% of connections on platforms like Facebook and X, formerly Twitter, are weak ties. This allows information to spread more widely on these platforms. There are several models for how information flows through social connections. Two such models are the independent cascade model (ICM) and the linear threshold model (LTM). They are variations of the basic information cascade model we already analysed.

In the independent cascade model (ICM), when a user "likes", comments, or otherwise interacts with certain content, they will be able to influence their ties with some probability. Imagine Alice sees a social media post and "likes" it. Bob, seeing that his friend Alice liked the post will also "like" or interact with it with some probability, let's say 40%. This probability will be different depending on certain factors, such as the content and the specific user. This process will continue through the network until some point when no additional "likes" or interaction will happen. As we have seen with information cascades, users may be more inclined to "like" a post that already has many "likes." In this rich gets richer scenario, the number of "likes" may be considered by users a more relevant factor than the content of the post itself. This is equivalent to the case of the hungry tourist who ignores her research and goes to the busiest diner.

The linear threshold model (LTM) defines a threshold per user. Only if this threshold is met, the user would take action. In this model, Alice may need to see a threshold of 10 friends liking a post before she does so. Another scenario where LTM may be suitable is when deciding to join a platform or a group. A user may want to see some of their friends act first. This is also a cascade model in which the actions of prior participants may be more relevant than the information the users collected on their own. Different models like ICM or LTM may be more suitable to different scenarios or platforms. But in all cases, we know that cascades play a very relevant role in information dissemination in social media and that weak ties help to increase the reach of each post. With that understanding, let's examine the role of social networks in information dissemination a bit further.

The digital-only myth "social networks"

There has been heated debate on the impact of social networks in society, ranging from the 2020 documentary *The Social Dilemma* to the more recent book *The Anxious Generation,* by Jonathan Haidt.[30] Although I personally agree with Haidt's views, my goal in this book is simply to analyze the effectiveness of social networks as an information dissemination system. I also understand the relevance of social networking platforms to connect communities, which I believe are both their strength and

---

30. *The Anxious Generation,* Jonathan Haidt, Penguin Press, 2024.

primary purpose. I'll not talk about that or other possible uses of social networks, focusing here only on their role as communication platforms.

There are a few points that make social networks ineffective in disseminating content:

1. They have a low barrier of entry. Unlike writing a book or a newspaper article, social networks make publishing videos, images, and texts very easy. That is one of the main features of these platforms. They need new content to be engaging and entice users to stay active. This is an asset if we are sharing recipes with our friends or posting the video of our kids' school recital. However, there are no explicit mechanisms, in any social media platform, to distinguish between innocuous posts and other content that may be considered news, including health care and political articles and videos.

2. They are prone to information cascades. Not only is the barrier for writing a post on social media platforms low, the barrier to interact with posted content, in the form of "likes," re-sharing, or comments is also low. It is much more convenient to click "like" in a post about a book or an article, than to read the book or article and write a full review. The creation of new hyperlinks to web pages, which involves editing the source code for the page, is also a much more time consuming endeavor than simply clicking "like." Users would only go through this trouble if they are really compelled

to add the hyperlink, as it cannot be done by simply scrolling a feed and clicking "like" by instinct. The low barrier of entry for interacting with content in social media platforms leads to herding, as we have seen in models like the independent cascade model (ICM).

3. They enable information to travel far. Most of the connections in social media platforms are weak ties, which enables information to propagate far beyond local communities throughout the wider network. Social platforms encourage the creation of weak ties, with features such as LinkedIn's "People You May Know" and platforms that display suggested content. These weak ties may be desirable in some scenarios, such as in job searches. In the case of information dissemination, they will amplify the effect of the information cascades.

These points are not my personal opinion of these platforms. They are a factual analysis of how their algorithms work. Although social networks may have positive impacts in some scenarios, they are an inefficient way to disseminate information. The low barrier of entry for posts, combined with the network-wide cascades make the system **fragile**. Basically, any content, regardless of validity, can become viral and rapidly spread through the network.

We all understand that when new technologies are introduced we need some time to evaluate their full effects. However, by now, the technical analysis I presented here shouldn't be a surprise, as everyone has experienced this in one way or another. Despite this awareness, the technical leaders of social networking platforms generally have not taken any meaningful action to make it either harder to disseminate misinformation or easier to verify facts in their platforms, and in fact, have taken measures to remove such barriers. In January of 2025 Meta, the parent company of Facebook, decided to drop its fake news filters, which marked posts as possibly fake.[31] They claimed that the quality of the filters were not good enough and that the platform could self-regulate their content. However, for reasons we describe here, we know it is not true. Fake news filters, while imperfect, helped increase the barrier of entry. Left on their own, Facebook users are bombarded with content promoted by information cascades, and have no mechanisms to sort such content or to impede low quality, misleading, or untrue content to become viral.

I like Yuval Noah Harari's definitions of myths, because we tend to forget that some of the systems we interact with on a daily basis and that are an integral part of our lives are just that. Only myths. This is certainly true for social networks, which are digital-only myths. There is a law demanding either the sale or the ban of TikTok operations in the U.S. because of its ownership by a Chinese company, ByteDance, and associated

---

31. https://www.nytimes.com/live/2025/01/07/business/meta-fact-checking (Last accessed in June 2025).

security risks. Such efforts recognize that a large number of Americans interact with content on the platform, which could manipulate such content against U.S. interests. Besides the TikTok ban in the US, in 2024, Brazil blocked the use of X for a few weeks, and Australia passed a law to ban social networks for children under 16. As these bans take effect, a few tech billionaires may become upset, but the world will go on. Gravity and the other laws of physics will not be impacted, and nothing will collapse. As technology leaders, it is our duty to actively investigate the positive and negative effects of our innovations, and to course correct when necessary.

How do we course correct such systems? It is imperative that we democratize publishing, like we did with the web. Besides lowering the barrier of entry and making content more available, we also improved latency. We can get news and information much faster than before. All that was possible while maintaining the **robustness** of the systems. Hyperlinks and algorithms like PageRank provided mechanisms to parse high quality, or at least highly endorsed, content from untrustworthy sources. By lowering the barrier of entry, we have quicker access to more news and publications.

Although social media lowered the barrier of entry even more by allowing more people to contribute more content, it did so using a fragile system that leads to **non-determinism** in the content it promotes. For example, we wouldn't go to a doctor that would always provide a wrong diagnosis. But somehow, we are fine with reading news from an unreliable source that we know is providing wrong results because its algorithms were designed to do so. With more and more reliance on online information and systems, the risks of lower digital literacy are significant. This is especially true for vulnerable populations, such as children and aging populations. We need to promote these debates and not accept the fact that a large part of the population is consuming news and information from fragile content dissemination platforms. With human agency, we can find practical solutions that can be implemented to limit social networks to scenarios that produce positive results, and balance such benefits with the harms of nondeterminism.

I chose to start our investigations of computer systems with social media because it is almost a textbook example of a technology that is impacting our lives, in many aspects negatively, by implementing fragile content dissemination algorithms that we don't care to understand. The more we understand how these systems work, the better equipped we'll be to demand and create change. In the same way we should investigate the qualifications of our doctor, we should also understand the systems that have such a huge impact in our lives. In the next chapter we'll continue to explore the problem of finding the most relevant information in the midst of more than one trillion web pages by going deeper into search systems.

3.

Search

Drew Barrymore, indices, COVID vaccines, scale, and online advertising

"Planet Earth has never been as tiny as it is now. It shrunk – relatively speaking of course – due to the quickening pulse of both physical and verbal communication."

Frigyes Karinthy, Chains

Brian Herzlinger was infatuated with Drew Barrymore since he was a little boy and saw her in the movie *E.T. the Extra-Terrestrial*. She was just six years old. Brian was a five year old in second grade at that time. He joined her fan club and had posters of her on his bedroom walls while growing up. Fast forward a few years, and Herzlinger was a broke 27-year old living in L.A. and working as an aspiring filmmaker. That's when he decided to try to get a date with Barrymore and document his quest in a movie.

His movie, *My Date with Drew* was released in 2004. It describes the whole story, from its ideation to the actual date, which happened after she found out about the project and was very touched by it. Parts of the documentary are a bit depressing, like when Herzlinger tries to get into shape in anticipation of the date. Or when he holds an audition to hire an actress that looks like Barrymore, and goes on a fake date with her, afraid that he wouldn't know how to behave if he did end up getting the date with the real Barrymore.

Excluding these parts, during most of the film Herzlinger and his friends try to explore the idea of six degrees of separation to get to Drew. For instance, they knew someone who was connected to John August, the screenwriter for *Charlie's Angels*, who worked with Drew in the movie. There were several three or four degree paths connecting Herzlinger to Barrymore. All of them proved useless. Hungarian author Frigyes Karinthy introduced the concept of six-degrees of separation in his 1929 short story *Chains*. The story describes a game in which participants try to find a connection to any other random person on earth, such as Selma Lagerlöf or an anonymous riveter at the Ford motor company, through only five other people, one of them an acquaintance.

Karinthy also indicates that this is a property of the modern world, which is striking considering that the story was published in 1929. This idea of connection is even more prevalent now, with connected devices and social media platforms, than in 1929. According to him, six-degrees wouldn't have been possible in the times of Julius Caesar, who would have had difficulties contacting anyone from the Mayan or Aztec empires. John Guare's 1990 play *Six Degrees of Separation* and the 1993 movie of the same title popularized the concept of six degrees of separation. There is even a game, *Six Degrees of Kevin Bacon*, where participants are challenged to link an arbitrary actor to the prolific Kevin Bacon in six degrees, in a chain of actors that collaborated in movies or commercials.

Although Herzlinger had many connections less than six degrees from Barrymore, that approach was not fruitful. He then created a website, which had a video explaining the project and a web page in which viewers could provide comments and suggestions. The website eventually became popular, with Herzlinger promoting it in radio shows and everywhere he could. Barrymore and her team found out about the project through the website, which led to her agreeing to go on a date and making the movie viable. Like Herzlinger, I too had a teenage crush on Drew Barrymore, but it never crossed my mind to try to find my connection to her using the six-degrees method.

When I was a teenager living in Brazil, it was likely that I could get to the country's president through my uncle, who had been a high-ranked official in the air force and later founded the Aeronautical Technology Institute (ITA, in Portuguese), one of the leading technology universities in Brazil. I can safely assume that the presidents of Brazil and the U.S. knew each other. I also think the U.S. president could have called Steven Spielberg, who knows Drew Barrymore. But since all these people are very busy and have more important things to do than connecting teenagers to their silver screen sweethearts, I don't regret that I didn't try to put that plan into action.

Facebook data confirms that, like Karinthy predicted, the connectedness of our modern world is making it feel smaller.[1] Its research team computed that the average distance between users in the platform decreased from 4.74 people in 2011 down to 4.57 people in 2016, when it had about 1.8 billion users, or 22% of the world's population, making the sixth degree concept closer to something like a digital fifth degree. As we have discussed in the previous chapter, most of these connections are weak ties, which behave as bridges connecting individuals, communities, and groups. In Malcolm Gladwell's view, "Six degrees of separation doesn't mean that everyone is linked to everyone else in just six steps. It means that a very small number of people are linked to everyone else in a few steps, and the rest of us are linked to the world through those special few."[2] In his view, one might assume that online connections make it even easier to connect with those special few individuals who have links to everyone else.

The concept of six degrees applies to more than people's connections. It is also applicable to concepts. The game *Six Degrees of Wikipedia* allows you to enter any two topics in Wikipedia to discover the distance between them.[3] And although I could devote a lot more time devising schemes to search for Drew Barrymore using my connections, it is more productive for us to move on

---

1. https://en.wikipedia.org/wiki/Six_degrees_of_separation (Last accessed in February 2025).
2. *The Tipping Point*, Malcolm Gladwell, Little Brown, 2000.
3. https://www.sixdegreesofwikipedia.com/ (Last accessed in February 2025).

to the topic of informational search. Let's start by taking a closer look at how book indices work.

Where are all the references to bubble sort?

A book index is very useful when you are trying to quickly review all the references to a given topic. In this book, I mentioned bubble sort a few times, including pages 8, 10, 44, 45, and 91 so far. These page numbers are illustrative and may not match the occurrences of "bubble sort" exactly depending on which version of the book you are reading. In the absence of an index, readers interested in going back to review some of these mentions would have to scan the entire book, which is both time consuming and error prone, as one can easily miss some important references if not paying close attention. In a digital version of the book, readers can simply search for "bubble sort" and review all the entries one by one. But with an average of 70 to 120 thousand words per book, what is the algorithm to efficiently enable users to search for a topic?

It starts with the construction of an index, which is very similar in spirit to those printed at the end of books. It is a collection of lists, one for each term. We normally refer to this type of index as an **inverted index**, as it is a mapping from terms to positions, "inverting" the content of the book. For example, the Constitution of the United States starts with "We the People," and the inverted representation would have a list containing the words "We," "the," and "People". The list for "We" would include all the occurrences in which that word appears in the Constitution, starting with its first occurrence at position 1. In another example, the inverted index for "to be or not to be" would look something like:

to: Positions 1, 5.
be: Positions 2, 6.
or: Position 3.
not: Position 4.

We can define and use positions differently. Position can mean the exact position in the text, as in this example, where with the precise position for every word in the index, we would be able to reconstruct the original text from the inverted index. This would be done by un-inverting the lists and placing the words in the right order. The word "to" goes to positions 1 and 5, "be" to positions 2 and 6, and so on. However, positions could be used to represent just the page number instead of the precise location, as in the traditional book indices. In this case, the inverted index is a lossy representation, as the original text can no longer be recovered from the index. Lossy representations are more compact representations that do not encode all the original information. In this case, with only page numbers represented, we wouldn't be able to reconstruct the original sentences.

We also have flexibility in defining what we use as terms in the index. Although "bubble sort" contains two distinct words, it functions as a single term referring to the algorithm's name. In the case of a book index, it is more useful for us to build an index of concepts and entities, than the full index of every word in the book. The full index with precise positions would be as large as the book itself and not very convenient to use. With the book index at their disposal, what would a reader have to do if they were interested in the C# description of the bubble sort algorithm shown in Figure 2? As the index entry for "bubble sort" is small, they could probably just go through every page listed in the index until they find what they want. If "C#" had its own index entry, it would be an even better choice because it includes fewer entries than "bubble sort." The two index entries, with occurrences up to this point in the book, are:

bubble sort: Positions 8, 10, 44, 45, 91.
C#: Positions 44, 45.

With the digital version of the book, however, a reader is able to pose a search query "bubble sort C#", which would return all the passages in which these two concepts "bubble sort" and "C#" appear together. In this case, these co-occurrences appearing in pages 44 and 45 would be shown. We can think about an efficient algorithm to do this. In the simplest case where the positions represent page numbers, finding the pages that mention two concepts requires us to find the intersection of the lists for these two concepts. In the search query "bubble sort C#", pages 44 and 45 are the only two that belong to the intersection, being present in both lists. The algorithm would start by looking at the search query "bubble sort C#" and checking if we have lists for these terms in the inverted index.

In this case we are lucky, as we have lists for "bubble sort" and for "C#". Identifying these lists seems simple enough at first glance, but there are some complexities here. For instance, someone could have misspelled the word "bubble." The search system would then identify that it doesn't have an entry in the inverted index for the misspelled word. It could either choose to return no results, or, if it is a bit smarter, it could identify by inspecting the lists in the inverted index that "buble sort" is close enough to "bubble sort," and that is probably what the reader wanted. In that case, it would notify the reader with a message like "returning results for **bubble sort** instead of **buble sort**," so they can double check that the automatic correction makes sense and rewrite the search query otherwise. Spell correction is one of the possible manipulations search systems can do to the user query, but there are many more. In some cases, search engines could expand the query adding alternative terms, including synonyms.

After we identify the two lists, for "bubble sort" and for "C#", we have to compute their intersection. We can assume that the lists are sorted, with page numbers increasing as we move through the lists. This will make the computer's job easier. The intersection algorithm will use a **cursor** per list. These cursors are nothing more than numbers that will track where we are in each list, like a bookmark that keeps track of which page you are reading. At first, all cursors point to the first element of each list, which are the pages that contain the first occurrence of each term in the book. In our example we have two cursors:

bubble sort: Positions 8, 10, 44, 45, 91. Cursor at 8 (the first position).
C#: Positions 44, 45. Cursor at 44 (the first position).

The next step is to check if both cursors are on the same page. If they are, we'd have a match, as we'd have found a page number that belongs to all lists, and therefore deserves to be added to the intersection set. In this case, as 8 and 44 are not equal, we don't have a match. Then, since the cursor for "C#" is at page 44, we know that there can be no match before that page. We can then safely advance the cursor for "bubble sort." And while the cursor for "bubble sort" is smaller than 44, or we reach the end of the list, we can keep advancing it. The first two steps would be:

bubble sort: Positions 8, 10, 44, 45, 91. Cursor at 8.
C#: Positions 44, 45. Cursor at 44.
No match, since 8 < 44. Advance cursor for "bubble sort." No results so far.

bubble sort: Positions 8, 10, 44, 45, 91. Cursor at 10.
C#: Positions 44, 45. Cursor at 44.
No match, since 10 < 44. Advance cursor for "bubble sort." No results so far.

We then get to the point that the cursors match on page 44, which is the first result added to the list. Once we find a match on page 44, we can advance both cursors. At that point we get the second match for page 45. We then try to advance the two cursors again, but since we reached the end of the list for "C#," we know there cannot be more elements in the intersection set and we are done. These steps would be:

bubble sort: Positions 8, 10, 44, 45, 91. Cursor at 44.
C#: Positions 44, 45. Cursor at 44.
Match, since 44 = 44. Add 44 to the result set and advance both cursors. Results: {44}.

bubble sort: Positions 8, 10, 44, 45, 91. Cursor at 45.
C#: Positions 44, 45. Cursor at 45.
Match, since 45 = 45. Add 24 to the result set and advance both cursors. Results: {44, 45}.

bubble sort: Positions 8, 10, 44, 45, 91. Cursor at 91.
C#: Positions 44, 45. Cursor at the end of list.
Done, since the cursor for "C#" reached the end of the list. Return the results: {44, 45}.

This is a simple enough algorithm. It is also efficient. Even if the list for "bubble sort" were really long, the algorithm would stop as soon as we reached the end of the smaller "C#" list. In the general case, a search query could have well more than two terms, some of them with possibly very long lists. The efficiency of the algorithm we discussed would still be proportional to the number of entries in the smallest of the lists. For instance, the word algorithm appears 115 times in this book so far. Efficiency appears 69 times and myth 54 times. The performance query "C# efficiency algorithm myth" would still be proportional to the small "C#" list. It is a bit surprising that in the presence of small lists, adding a very large list does not degrade the performance of the search query.

Frequently occurring words, like "the," which appears 6,487 times in this book, end up not impacting the performance of search query at all, as the algorithm will always be driven by the smaller lists. These recurring terms are often called **stopwords**. They are normally considered irrelevant for search systems, and are many times disregarded completely. For instance, every web page has the term "www" so adding "www" as a search term doesn't help the search engine to filter out any pages and the list can easily be disregarded. The presence of stopwords, like "www" in this case, won't impact the results. Let's consider another example with two lists:

the: Positions 1, 2, 3, 4, 5, 6, 7,... 300. Cursor at 1.
rare term: Positions 57, 170. Cursor at 57.

Since the first possible hit is at page 57, we'd advance "the"'s cursor until page 57. Then we'd add 57 to the result, and move both cursors to the next document, as shown below:

the: Positions 1, 2, 3, 4, 5, 6, 7,... 300. Cursor between pages 1 and 56.
rare term: Positions 57, 170. Cursor at 57.
No match, while "the"'s cursor < 57, advance it to the next page. No results so far.

the: Positions 1, 2, 3, 4, 5, 6, 7,... 300. Cursor at 57.
rare term: Positions 57, 170. Cursor at 57.
Match, since 57 = 57. Add 57 to the result set and advance both cursors. Results: {57}.

the: Positions 1, 2, 3, 4, 5, 6, 7,... 300. Cursor at 58.
rare term: Positions 57, 170. Cursor at 170.
No match, while "the"'s cursor < 170, advance it to the next page. Results: {57}.

As 170 is also a match we would add it to the result set and be done, as we'd reach the end of "rare term"'s list.

the: Positions 1, 2, 3, 4, 5, 6, 7,... 300. Cursor at 170.
rare term: Positions 57, 170. Cursor at 170.
Match, since 170 = 170. Add 170 to the result set and advance both cursors. Results: {57, 170}.

the: Positions 1, 2, 3, 4, 5, 6, 7,... 300. Cursor at 171.
rare term: Positions 57, 170. Cursor at the end of the list.
Done, since the cursor for "rare term" reached the end of the list. Return the results: {57, 170}.

```
1  static int[] FindIntersection(int[] array1, int[] array2)
2  {
3      List<int> intersection = new List<int>();
4      int i = 0, j = 0;
5
6      while (i < array1.Length && j < array2.Length)
7      {
8              if (array1[i] < array2[j])
9              {
10                  i++; // Move cursor in array1
11             }
12             else if (array1[i] > array2[j])
13             {
14                  j++; // Move cursor in array2
15             }
16             else
17             {
18                  // Found a common element
19                  intersection.Add(array1[i]);
20                  i++;
21                  j++;
22             }
23     }
24     return intersection.ToArray();
25 }
```

Figure 5. Intersection algorithm described in C#. Variables i and j are the indices of the two cursors for the lists in *array1* and *array2*, respectively. The output is stored in variable *intersection*. Line 19 adds the match that we found, when *array1[i]* is equal to *array2[j]*, to the output list.

Figure 5 shows the C# representation for this list intersection algorithm.[4] Again, there is no need to understand the details shown in Figure 5, it is only for illustration. In this version of the algorithm, we use integers to represent each item's position in the list. However, as in real search engines these lists tend to be very large, positions are typically stored in compressed format. There are several ways to efficiently compress these lists, using approaches similar to the Huffman coding we discussed in Chapter 2. Additionally, there are techniques to speed up advancing cursors to a given position. For instance, when "the"'s cursor is at position 58 and we know that the next possible match is only at position 170, we can use these techniques to efficiently advance it 170. This allows the time spent by computers to do the intersection to be proportional to the smallest list.

But what can we do if we want to find results where "C#" and "bubble sort" appear in the same paragraph, or the same sentence? We can change what we store in positions, from page numbers to more precise location references. One approach is the use of full occurrence positions, such as "We" is in position 1, "the" in 2, "People" in 3, and so on. In this case, each word in a book would have its own position and two words would never belong to the same position. This would require changing the intersection algorithm above to instead consider a match

---

4. Please see a visualization of the list intersection algorithm at https://digitalagencybook.org/visualizations/list-intersection.

when the positions are within a certain distance from each other. We might use a distance of 15 to 20 positions to align with the average length of a sentence in English.

Another option is to use the Dewey decimal system. The Dewey system was originally devised to organize libraries, but we can also use it to represent positions. Using Dewey, we can represent the first chapter as 1. The first section of chapter one would be 1.1. The fifth sentence, of the second paragraph, of the first section, of the first chapter would be 1.1.2.5. This Dewey-based representation encodes more information into positions. This allows queries to be more precise, enabling finding matches at the chapter, section, paragraph, or sentence level, at the expense of being more complex to represent and compress. Before we add more details to the inter-section algorithm, let's talk more about search in physical libraries.

Organizing libraries

Before standardized classification systems were widely adopted, libraries assigned book positions based on factors like height or date of acquisition.[5] This wasn't a problem initially because only a small number of privileged patrons could access the collection, and they were often assisted by librarians who knew the system. During the 20th century, however, this changed as library

---

5. https://en.wikipedia.org/wiki/Dewey_Decimal_Classification (Last accessed in February 2025).

access was democratized. With more patrons wandering through the shelves, a classification based on the relative similarity of topics made more sense. This need for structure inspired American librarian Melvil Dewey to define the Dewey decimal system. There have been multiple revisions of the classification, but the basic design remains the same.

There are ten classes, each divided into ten divisions, each having ten sections. For instance, class eight refers to literature. 81 is the division for American literature in English. Section 813 refers to American fiction. The ten main classes are the following:

000 – Computer science, information and general works
100 – Philosophy and psychology
200 – Religion
300 – Social sciences
400 – Language
500 – Pure science
600 – Technology
700 – Arts and recreation
800 – Literature
900 – History and geography

The three initial numbers can be extended with decimals for further specificity. For instance, the primary Dewey for this book is 004.019. In its first three digits, the first 0 covers general knowledge, reference, and information systems, second 0 indicates generalities, including large-scale works about knowledge systems, computing, and organizations of knowledge, and 4 refers to data processing and computer science, composing the integer part of the Dewey number, 004. In the last 3 digits, decimal 019 denotes social and ethical issues in computer science. This classification scheme is an efficient solution for organizing books by topic. To look for broader topics, users and librarians are free to ignore the rightmost digits of the Dewey number and focus only on the higher level classifications.

This hierarchical organization also holds when we use Dewey numbers to represent index positions. Let's say the position components represent chapters, sections, paragraphs, and sentences. By using the full Dewey number we now can find matches in the same sentence. If we ignore the rightmost digit we can find matches belonging to the same paragraph, and ignoring the last two digits we can find matches belonging to the same section. For instance, 1.1.2.7 and 1.1.5.2 could be considered a match if we are interested in occurrences on the same section, regardless if they are on the same paragraph or sentence.

When a library is organized using the Dewey system, a patron can find shelves holding books of similar topics, such as South American politics, through a computer system or by asking a librarian. This scheme is way more efficient when we are searching for books than the original organization in which books were sorted by height or acquisition date. Considering that there are 2.5 million books in the New York Public Library,[6] organization by Dewey has proven very useful. Even if a patron had full access to all its shelves, it would be incredibly inefficient to find the appropriate books when researching a topic without an organization system.

Before the internet and the digitization of libraries, finding references when researching a topic was a burden. Researchers had limited access to information, and sometimes only a few libraries in the world carried some manuscripts. Even when I was a graduate student in the nineties, researching all the published bibliography about a topic, such as compression algorithms, was so laborious that in a few cases it warranted a research paper in itself or even a master thesis. In the AI era, most chatbots can produce very thorough research about any field without the need for time consuming, and sometimes expensive, trips to libraries.

Searching for books was so difficult that in 1983 the United Kingdom Yellow Pages ran an advertisement in which an older gentleman used the Yellow Pages to locate a copy of a hard to find book. He used the Yellow Pages to call local bookstores. After several attempts, he finally found a bookstore that had the book and succeeded in securing a copy. He identified himself as J. R. Hartley, the book's author, who was searching for a copy of his own

---

6. https://www.nytimes.com/2015/11/28/arts/design/a-slippery-number-how-many-books-can-fit-in-the-new-york-public-library.html (Last accessed in February 2025).

book. The ad was promoting the effectiveness of Yellow Pages to find obscure items. And it became so popular that several U.K. bookstores started receiving requests for the fictitious book *Fly Fishing*.

That prompted Michael Russell, who had written a book about fly fishing, to republish his book under the name of J. H. Hartley. *Fly Fishing: Memories of Angling Days*, was published in 1991. The actor that appeared as Hartley in the television commercials was hired to promote the book as the public face of the author. This marketing stunt was very successful. The republished book sold 130,000 copies during the Christmas season and made the bestseller list in the United Kingdom in 1991.[7] This story would not have happened today, when all published books are easily accessible online. To me, despite all the convenience of buying a book with one click, there are some human and even nostalgic elements in how J. R. Hartley found a copy of his own book. Calling bookstores was fun, and that's what attracted people to the Yellow Pages ad. But let's set nostalgia aside and go back to search. During the early days of the COVID-19 pandemic, the frantic search for COVID vaccines in the U.S. echoed J. R. Hartley's painstaking search for his own book.

---

7. https://en.wikipedia.org/wiki/Fly_Fishing:_Memories_of_Angling_Days (Last accessed in February 2025).

Finding a COVID shot

Amid global crisis and public health emergencies, COVID vaccines were developed and authorized at an unprecedented pace of less than a year. This was considered **warp-speed,** given that usually vaccines take ten to fifteen years to be developed. In late 2020, after the vaccines were authorized, came the second problem: how to efficiently distribute them to the population? The U.S. used a phased rollout. Phase 1A, from December 2020 to January 2021 was limited to health care workers and residents and employees of long-term care facilities. Subsequent phases included essential frontline workers, people with some underlying health conditions, and people above certain age restrictions, such as 75 and other and subsequently 65 to 75. After the rollout to these target groups, the vaccines became more broadly available to the general population. This prioritization was advised by Centers for Disease Control and Prevention (CDC) and the Advisory Committee on Immunization Practices (ACIP), while each state was responsible for the actual distributions, following guidelines from the federal government.

There were supply chain complexities for vaccine distribution. For instance, the Pfizer vaccine required ultra-cold storage. Additionally, all the misinformation and hesitancy about COVID itself and the vaccines posed real challenges for equitable and efficient distribution of the vaccines to the broader population. In most cases, states worked with pharmacies, hospitals and community centers to administer the shots. Despite the program's overall success, the experience for most people in the U.S. was inconvenient.

There was no central system that listed where vaccines were available and allowed people to schedule an appointment. Instead, most websites listed the places in your vicinity that were administering shots and you had to contact each of them individually to try to schedule vaccination appointments. Some providers offered scheduling online, while others required phone calls. I still remember the frustrating weeks I spent regularly checking the websites for all vaccination clinics in my vicinity to schedule appointments for my family and I.

Perhaps due to the combination of all these reasons, the percentage of vaccinated individuals also varied widely across states.[8] By January of 2021, North Dakota and West Virginia had the highest percentage of vaccines distributed by population, 90% and 85% respectively. In contrast, Massachusetts ranked 32nd of U.S. states with only 5.2% of the population having received the first dose of the vaccine. Massachusetts eventually caught up, jumping to 16th place by the end of February. This uneven demand for vaccines across states made efficient distribution of vaccines even harder.

---

8. Why Have Some States Been More Successful at COVID-19 Vaccine Distribution than Others?, G. Wilensky, Milbank Quarterly Opinion, 2021. Available at https://doi.org/10.1599/mqop.2021.0415 (Last accessed in February 2025).

If the vaccine inventory had been consolidated in a single search system, citizens would have been able to book their appointments more easily, possibly even with just a single search. However, with no consolidated inventory and no central search system, we had to rely on the different websites for all the different public and private vaccine administration sites. Every day, likely multiple times a day, individuals had to search across various systems belonging to CVS, Walgreens, and others, in case a new batch of shots arrived and new appointments became available.

The ideal system would have provided a single point of contact for individuals to register their families, providing their ages and relevant information that may impact their vaccine priority. Based on this information, the system would know the demand for vaccines in each area and could distribute them accordingly. The system could even notify individuals of their upcoming appointment dates, as subject to vaccine demand and availability. Having the exact data of how many individuals, by county, wanted to be vaccinated and were eligible, could have streamlined distribution, especially due to the uneven demand for shots. Ideally, we'd brainstormed and devised an easier to use and more equitable distribution system, perhaps even more efficient to what I'm proposing here, and this system would have been implemented and available even before the vaccines had been approved by the U.S. Food and Drug Administration (FDA).

Let's say we succeeded at scheduling a vaccine appointment and went to a pharmacy to get our shot. As anyone who has been to a busy pharmacy knows, with too many patrons and too few cashiers, we may get very long lines and wait times. Having too many cashiers can be good for customers, but it is wasteful. An efficient system would have just enough cashiers to not let customers wait too long. Then there is the problem of how to structure the lines. In COVID's case, states controlled their own distribution. And in many cases, that was delegated down to the pharmacies and distribution centers, who each managed their "lines," to assign vaccine appointments.

Single-line systems are more efficient and provide more predictable wait times, but they have some downsides. Users see a single line that may appear very long. They also may feel a lack of control, since they can't choose the shortest line. And a single line, in some situations, may require more physical space, being difficult to implement. Some pharmacies and supermarkets implement a hybrid approach, with a single line for self-checkout and multiple lines for the cashiers.

One possible metric of efficiency for line systems is **utilization**, computed using arrival rate of requests and the time to serve each request:

Utilization = Arrival Rate / Time to Serve

In a perfectly balanced system, when the arrival rate and the time to serve are equal, the system is 100%

utilized. If the arrival rate is two times faster than the processing time, the system would be oversubscribed with 200% utilization and customers waiting in line. An oversubscribed system is one in which we have more demand than availability of resources, as when airlines sell more tickets than the number of available seats in a plane. In the opposite situation, when the processing time is twice faster than the arrival rate, the system is only 50% utilized and the cashier is idle half of the time.[9]

With just one line, the arrival rate would be the aggregate arrival rate of the entire population. This would be all the people in the U.S. wanting shots or all the patrons in the store. With multiple lines, such as one per state, with varying arrival rates, some lines can achieve higher utilization over others. This is what we saw empirically during COVID, given the uneven demand for vaccines. When I was searching for appointments in the multiple vaccination clinics near me, I was in fact trying to find a "line" with lower utilization to schedule the shots for my family and I. A single-line system would have solved the problem of constantly checking different providers for vaccine availability.

One reason for using multiple lines is scale. There are pros and cons to scaling a system. A system that is operating below its expected utilization is wasteful. You still have to pay the cashier's hourly pay even if the system

_____

9. Please see a visualization of this cashier line system at https://digitalagencybook.org/visualizations/cashier-lines.

is 50% idle and the cashier is busy only 50% of the time. On the other hand, in an oversubscribed system where there is an excess of customers waiting in line, it may be possible to add extra cashiers and at an incremental cost. The utilization of the system is a good metric that allows us to reason about the number of cashiers. If we start with an oversubscribed system and keep adding cashiers, eventually we'll reach a point when adding more cashiers won't help. At this point, the efficiency of the system is **optimal** and it is perfectly balanced, with arrival rate equal to the time to serve. In an optimal system, utilization is 100% and we don't need to add more cashiers. The popular term **economies of scale** refers to the economical analysis of the cost advantages of scaling a system. With that in mind, we'll go back to explore how scale impacts the use of indices for search queries.

Scalable search systems

The search index for a single book is small and can be searched on a single computer. That is no longer true when we talk about the more than one trillion web pages available on the internet. For web search engines, such as Bing and Google, there are two main scalability factors: the number of web pages to index and the volume of search queries to answer per second. Let's start with the large number of web pages in the search index.

When we talked about a book index, we used positions to represent page numbers. In the case of web search, they represent web pages. For instance, we can assign positions as:

Position 1: https://en.wikipedia.org
Position 2: https://en.wikipedia.org/wiki/
Drew_Barrymore
Position 3: https://www.microsoft.com
Position 4: https://www.bing.com
…
Position one trillion: https://www.fontoura.org

Most search engines also store the term position within the page in addition to the page identifiers. For instance, since the first two words in Drew Barrymore's Wikipedia page are respectively Drew and Barrymore, they would have positions 2.1 and 2.2. Position 2 is the page itself while the fractional number represents the occurrence of the word within the page. Assuming we need two numbers per position, one for the page identifier and the other for the term position within the page, we'd need about 2TB, or two trillion bytes, to store a list with one trillion positions, if these numbers were efficiently compressed.

It is likely that stopwords like "the" and "www" will appear more than once per page, so we could end up with a few lists even larger than 2TB. The scale comes from the fact that we must store in the inverted index one entry for every word in every web page. Each of these entries will belong to one of the lists. Of course, the vast majority of lists will have way less than one trillion entries, as most of the words are not that popular and don't appear in very many pages, much less in every web page.

There are over a million words in English alone and over seven thousand languages in the world. If we add up the sizes of every list that should be indexed, it would be way larger than what we can store in a single computer. A **server** is just a powerful computer designed to provide data or services to other computers over a network. The average memory size of today's servers is smaller than 2TB, so in a single server we cannot even store the list for stopword "www" in memory. Once we realize that we need more than one server to store the full index, we have a few options to divide the index across multiple servers.

The first option is to split lists between servers. For instance, if we had ten servers, and lists for ten thousand words, we could say that each server would store the lists for one thousand words:

Server$_1$: Lists for word$_1$... word$_{1000}$
Server$_2$: Lists for word$_{1001}$... word$_{2000}$
...
Server$_{10}$: Lists for word$_{9001}$... word$_{10000}$

With this organization, to answer a search query containing two words like "Drew Barrymore," we need to identify the appropriate servers that contain the words' lists for each of the query terms, and retrieve these lists to perform the intersection algorithm. The list for "Drew" may be in Server$_2$ while the list for "Barrymore" may be in Server$_7$. We'd need to ship these lists to a single server, preferably either Server$_2$ or Server$_7$ to minimize the amount of information that needs to be shipped, to perform the intersection algorithm and return the results. There are several techniques for distributing the lists across servers. A simple way to do it would be alphabetically. A more sophisticated approach would be to sort the terms by the likelihood that they appear in search queries. Distributing terms across servers by their propensity to be searched would avoid a single server having all the popular terms, and being too busy, while other servers would be mostly idle.

Regardless of the way we distribute the terms, it is very likely that we'll need to fetch lists from different servers when a search query has two or more terms. As the number of search terms in a query increases, the number of servers involved typically increases as well. After we retrieve the lists and ship them to a single server, we will compute their intersection using the same algorithm described in Figure 5. The only difference now is that the results correspond to web page identifiers instead of pages within the book. We call this organization that splits the inverted index by terms, **term sharding**.

The other popular method of handling scale is splitting the web pages instead of splitting the terms. Whereas before we were storing complete lists of locations for a given word, now we're storing partial lists of locations for a given word on every server. Partial here meaning that we'll only list the locations of the word that are covered by the server's dedicated pages. In other words, in this new organization, all the servers would have all the lists. However, they would have only a subset of the web pages, so their lists would be smaller. With ten servers, the lists would be ten times smaller. Considering we have ten thousand pages, each of the ten servers would have lists for about one thousand web pages:

$Server_1$: Lists for every word, containing only locations for $page_1 \ldots page_{1000}$

$Server_2$: Lists for every word, containing only locations for $page_{1001} \ldots page_{2000}$

...

$Server_{10}$: Lists for every word, containing only locations for $page_{9001} \ldots page_{10000}$

With this organization, every server is involved in every search. For the query "Drew Barrymore," each server would perform the intersection algorithm in the subset of documents it contains and ship its partial results to a single server, which would then consolidate the ten partial results to produce the final list of matching web pages. We call this organization that splits the inverted index by documents, **document sharding**.

The number of lists and the sizes of the lists being intersected don't change if we split by terms or by documents, but both of these methods have pros and cons. Term sharding is generally better for search queries involving small lists. As we need to have all the lists in a single server to perform the intersection, if some of the lists are large we'd have to ship large amounts of information to get all the lists to a single server. Document sharding minimizes the need to ship lists around and allows for a more even distribution of the computation amongst the servers. Every server participates in every query. This is a desirable property as the total computational resources required to answer the search query will be evenly distributed amongst all the servers.

In the same way that we like to see lines with equal length in a pharmacy or supermarket, a good property for computer systems is to maintain, as much as possible, even utilization across the servers. If one of the supermarket lines is longer than the others, it is likely that the cashier is slower than his peers. He may also be super friendly and customers don't mind waiting a few more minutes just to talk to him. In the case of web search, however, the imbalances are most likely due to uneven demands, such as one server having to intersect a very large list or ship it to another server in the case of term sharding. Most large-scale search engines use document sharding, as that allows the intersection computation involving large lists to be evenly distributed amongst the servers. There are also hybrid schemes, which combine the gains from both document and term sharding schemes, but are more complex to implement.

The first reason to **shard** the inverted index is its sheer size, which can easily extend beyond the limits of a single server, resulting in a need for splitting information across multiple servers. The second reason is the huge volume of queries web search engines receive every second. When I asked Bing how many search queries it processes per second, it returned 10,417. Even discounting my search, 10,416 is still a lot.[10] In a typical 3GHz computer, with the simplistic assumption of one instruction per cycle, we would be able to perform 3 billion instructions in one second. This would allow approximately 300,000 instructions for each of the 10,000 queries needed by Bing in a second. As we have seen, there are lists with over 300,000 entries. Therefore, we'd likely not be able to intersect these large lists within this time budget.

Search volume is so severe for the popular search engines that it has become the determining factor in calculating the number of servers needed in their deployments. To meet user demand, search engines need more servers than the minimum required to store the inverted index. The exact number is computed by the processing capacity of the servers and the aggregate number of instructions required to compute all the simultaneous 10,416 searches every second. Most of these instructions will be used for decompressing and intersecting lists. It is a boring task, but computers don't mind the repetitive work.

---

10. This is a joke, as the returned answer is likely a precomputed average across some time period, and therefore my search couldn't possibly influence the result.

Another reason that search engines must use more than one server is **resiliency**. Servers may fail and having extra ones as backup is often a good plan. It's just like how having an extra car in your garage in case your primary car has issues can add resilience to your busy schedule. However, resiliency can also be very costly. Buying an extra car to leave it unused most of the time may not be wise, as it is very inefficient. Similarly, to balance **efficiency** with **resiliency**, these extra backup servers should ideally not be left idle while waiting for the primary server to fail. They are typically also actively helping users search the web. With more than enough servers to satisfy the high query load, when one fails, the others continue working without disrupting user queries. Having these active backups is like having a household with the right number of cars such that every car gets good usage and, when one is being serviced, there are still enough cars for everyone to get to work and school on time.

Despite their significant advantages, both term and document sharding are not bullet proof solutions. If we use document sharding, for instance, when a server fails we still lose the results for the queries already in-flight. This means that even though the remaining results are still loaded from the working servers, the quality of the answers returned to the users might suffer. In our example with the ten servers and ten thousand web pages, if $Server_2$ fails we'd lose the results from $page_{1001}$ to $page_{2000}$, but still have the full results coming from the other nine thousand pages. We could be very unlucky and miss an important result, such as the Wikipedia entry for search "Drew Barrymore," but search engines implement mechanisms to mitigate that, such as storing multiple copies of important pages in the search index.

Resiliency is always a consideration in computer systems, as machines can fail in unexpected ways. Whenever we have a **distributed system**, involving more than one machine, the failure patterns start getting more complicated. Leslie Lamport, winner of the Turing Award, which is the highest honor for computer science researchers and often likened to the Nobel Prize, defines a distributed system as "one in which the failure of a computer you didn't even know existed can render your own computer unusable." Explicitly thinking about how to make systems more resilient is key. As we discussed before, a fragile system that doesn't properly work is by definition an inefficient system. A sorting algorithm that returns a wrong result very quickly is not efficient. Similarly, a fragile web search system would leave users frustrated, as their queries would take too long or never get answered. These users would eventually stop using the system, and the entire investment we made would go to waste. Fragility leads to inefficiency.

All large distributed systems will have to deal with failures. As we've seen, black swan events are to be expected. On August 8th, 2022, for instance, a Google data center in Iowa caught on fire. In addition, another, unrelated problem with a software update on the same day, caused many of its services to fail, including Google search, maps, and YouTube. These types of problems happen rarely, especially in companies such as Google, which invest heavily in making their systems reliable. That day they had two problems. Sometimes when it rains, it pours. This example highlights that companies that invest in resiliency for their systems can minimize the impact of failures, but will never be completely free from black swan events. They are a fact of nature and companies need to learn how to deal with them smoothly when they happen.

When talking about this point with my colleague Michael Schwarz, who is Microsoft's chief economist, he made this very interesting comment: "It is interesting how economics and computer science ways of thinking at times are identical and at times diametrically opposite. An economist would expect that 'these kinds of problems' happen more frequently in companies that invest heavily in making their systems reliable. This is because economists are obsessed with **endogeneity**. Hence, I would expect that companies that, due to the nature of their business, suffer the most from 'these kinds of problems' would invest the most into making their systems reliable to at least partially mitigate the problem. Another example of similar reasoning: students with perfect SAT scores are, on average, probably NOT the students who studied the most for the test, because they did not need to."

Endogeneity means that some of the factors may be hidden, such as the innate smartness of students, possibly making some conclusion unreliable, like the students that studied the most didn't get the best grades. I get his point, and I completely agree with the example of high SAT scores not coming from the students who study the most. However, as we'll see throughout the book, computer systems are so interconnected today that every company with an online presence or offering online services is susceptible to black swan events, not only the giants big techs like Google. We'll talk more about resiliency in Chapter 4, when discussing cloud computing.

Next, we'll explore how PageRank, which helps search engines distill the high quality pages from the lower quality ones, works alongside the list intersection algorithm to increase the relevance of search query results.

Returning relevant results

The main goal of a search engine is to return the most relevant results for a user query, in sorted order, with the most relevant web page on top. This is done in two phases. In the first phase, the intersection algorithm we discussed so far first produces all the results that are viable answers to the query. If the query is "Drew Barrymore," the viable results must contain these two words. There are exceptions. For queries involving rare terms, search engines may decide to drop terms to generate enough results, but these exceptions are rare. In most cases all words in the search query should be present.

In the second phase, after the intersection algorithm produces the viable results, we must sort these results and select the top ones to present to the user, in order of their relevance to the query. There are two main factors that go into computing how relevant a web page is for a given search query: the overall quality of the web page itself and its relatedness to the search terms. The first component is independent of the search query, and although there are other factors that may go into it, with each search engine having its own secret sauce, we can assume it is proportional to the web page's PageRank value. We can write the score of a page P to a search query Q as:

$$\text{Score}(P, Q) = \text{PageRank}(P) + \text{Similarity}(P, Q)$$

The second part of the equation, the similarity score, measures the prominence of search terms on the page. For example, if there are many occurrences of the words "Drew" and "Barrymore," it is considered good. If these occurrences appear together in the page, it is even better. If the words appear in headers or in the page's title, it may be an even stronger signal of their relevance to the page. To compute this similarity score, search engines use each term's positions within the page. They may also use extra information that indicates the **salience** of the term occurrence, such as whether it is in bold or italic, or if it is in a header. For that, of course, they would have to store extra information in the lists, in addition to the positions. The more information they store, the more precise the similarity score gets at the expense of the lists taking up more space in the inverted index. As with PageRank, each search engine has its own secret sauce for computing similarity.

Although logically we have two phases for returning search results, intersection and computation of *Score(P, Q)*, these are normally done at the same time for efficiency's sake. Web search engines compute the PageRanks daily and store them in a list. As the intersection happens, we compute *Score(P, Q)* for each page *P* deemed as a match. This computation combines the prestored PageRank value for *P* with *Similarity(P, Q)*, computed during the intersection using the positions and salience information stored in the inverted index. For most popular queries there are literally millions of potential candidates. Search engines select the top documents by *Score(P, Q)* and discard most matches.

So far, we have used hyperlinks only in the computation of PageRank, the query independent part of the score. However, there is also a smart way in which links can influence the relevance of a page to a user query. **Anchor text** is the text surrounding a link. Consider the snippet "Drew Barrymore is the most beautiful and talented actress in the world," in which Drew Barrymore is a link to her Wikipedia page. The whole sentence would be the **anchor text**. Every link to web page *P* has its associated anchor text. We can then compute the aggregate of all these anchor texts. For instance:

Page 100 → Page 2: Drew Barrymore is the most beautiful and talented actress in the world

Page 581 → Page 2: The gorgeous actress Drew Barrymore
Page 11281 → Page 2: Charlie's Angels star, the goddess Drew Barrymore

...

A somewhat surprising finding was that aggregate collection of anchor text is a very accurate description of what the target page is about, and often a better indicator than the text of the page itself. Take this example, what could possibly describe Drew Barrymore better? What makes anchor text so good is that it is a summary, by the authors of the source pages, of what the content of a given target web page is about. Once researchers understood anchor text was effective, search engines wanted to incorporate it somehow in the score calculation, boosting the value of *Similarity(P, Q)*.

There are different ways search engines may choose to do that. A natural way is to augment the content of a page with its anchor text. In this example we'd add the sentences "Drew Barrymore is the most beautiful and talented actress in the world. The gorgeous actress Drew Barrymore. Charlie's Angels star, the goddess Drew Barrymore" as an appendix to Drew's Wikipedia page. This addition is not a real modification of the page's content, but the inclusion of the sentences coming from anchor text to the inverted index. The list for word "goddess" would then have an entry for position 2, which is the identifier of the target web page, even though "goddess" may not appear in that page itself. The computation of *Similarity(P, Q)*, knowing that anchor text is more accurate than the content from the page itself, may use this information to boost the score for page 2 for queries containing the term "goddess."

Web search engines also use the feedback from users to refine their score computations. This technique is called **relevance feedback**. The most effective signal users can provide are clicks. When users click on a result, we have positive feedback. When users skip results and click on a link further down the search results page, we have negative feedback for the skipped results and positive feedback for the one that was clicked. There are also more complex signals, such as dwell time, which indicates how long the user stayed in the clicked page before going back to explore the other search results. All these signals are factored into the score computation.

There is no direct way that website owners can influence the placement of a page for a given search query. It all depends on its authoritative value, captured by the PageRank score and anchor text, and how well the page matches the query. None of these can be directly manipulated. It is possible to try to fraudulently boost a page placement, by creating fake links and anchor text, or by hiring an army of robots to click on search results. These tend to be ineffective, as search engines have entire departments dedicated to combating fraud. As discussed earlier in the book, search results are very stable and not susceptible to herding. This makes organic search engine results trustworthy. Next, we'll talk about online advertising and show that, unlike search, it allows advertisers to directly manipulate the placement of their ads.

The trouble is I don't know which half

John Wanamaker, a nineteenth century retailer, is famous for a quote that is often attributed to him: "half the money I spend on advertising is wasted; the trouble is I don't know which half." The advertising market is huge. It surpassed over one trillion dollars in 2024.[11] Wanamaker was right that it is hard to quantify the benefits of advertising, especially before most of it became digital. Currently, 70% of global ad spend is online. In the US, this number is 80%, reaching 270 billion dollars in 2023. Most of that spending goes to big tech companies to place advertisements on their platforms.

Compared to placing an ad on a magazine, radio, newspaper, or TV, online advertising provides several advantages. Two of the biggest ones are fine grained targeting and better performance metrics. Wanamaker would be pleased to know that the online advertising campaigns can target specific user segments and we can measure the relative performance of these campaigns. Whereas placing an ad in Rolling Stone once aimed broadly at reaching a young, hip audience, today's advertisers can target that same demographic, and much more

---

11. https://www.ft.com/content/e9d9befb-d5fd-438e-89d3-47f894c56736 (Last accessed in February 2025).

precisely, through online platforms. Instead of guessing who might flip through a magazine and casting a wide net, tech companies now use detailed user data, including demographics, browsing behavior, and brand interactions, to deliver ads directly to the most relevant individuals.

It is also difficult to know the effectiveness of ads placed on the physical version of Rolling Stone. Have readers from the target audience even seen the ad? If so, how many views actually resulted in action? Ads on rollingstone.com, on the other hand, are tracked. We know many users have seen each ad and which demographics responded better to each advertising campaign. In several cases, it is even possible to track whether a given ad view led to a sale, enabling advertisers to measure their return on investment. There are many types of online advertising, including search, social media, and video.

Search advertising is the most popular form of online advertising. In the U.S., it represents 40% of total advertising spend, being forecasted to reach $154 billion in 2025.[12] This is gradually changing, with the rise of social media and with new AI tools, but search advertising still captures a very large market. The heart of search advertising is that it essentially piggybacks on web search, even leveraging the same interface. Ads are typically placed on the top of the search results page, before the web search results section starts. The ad results are explicitly marked

---

12. https://www.statista.com/outlook/amo/advertising/search-advertising/united-states (Last accessed in February 2025).

so users can distinguish between them and the results returned from the web. Similarly to the web results, the displayed ads are typically related to the user query. Unlike web search, search engines don't need to return ads for every user query. Some user queries may not be of commercial nature, such as the search for a research paper. For instance, the query "page rank paper" in Bing does not return ads.

Ads and search results are selected by entirely separate systems, but their results are combined to form the search results page you're familiar with today. There are also other systems that feed results into the search results page, including AI-generated answers, maps, and summaries of Wikipedia entries. Not every query displays results from all these systems, also called **backends**. After all the backends return their results, web search engines have a dedicated system to combine them into a cohesive page. When I worked on ad selection at Yahoo, one of our goals was returning ads that were consistent with the rest of the page.

One interesting query we encountered was "latin canon." It is likely that a user searching "latin canon" is looking for the Canon of the Mass, a traditional Catholic prayer.[13] While the organic search results reflected that intent, most of the ads assumed the user was in the market for a Canon printer. We made several improvements to Yahoo's ad

---

13. https://en.wikipedia.org/wiki/Canon_of_the_Mass (Last accessed in February 2025).

section systems to mitigate this problem. For instance, we worked on automatic mechanisms to identify that queries like "latin canon" are not commercial, and therefore should display no ads. We also provided mechanisms to guarantee that ads and search results, despite being selected by different systems that were unaware of each other, were combined consistently into a cohesive search results page.

Our group at Yahoo Research, led by Andrei Broder, who is now a Distinguished Scientist at Google, pioneered the field of computational advertising. Our key contribution was to treat the problem of selecting ads as a search problem. Ads results should be relevant to the user query and consistent with the results from the web and the other backends. This is very hard to do, given the distinct nature of ad selection:

- While there are trillions of web pages, the number of active ads campaigns is much smaller.
- Web pages typically contain substantial information about a subject, while ad descriptions tend to be short and provide less content to the selection algorithms.
- Web page links allow algorithms like PageRank to distill authoritative pages from less quality ones. It also provides anchor text, which are useful descriptions of the pages' contents. Ads don't have hyperlinks.

The work in computational advertising strives to select relevant ads to search queries, despite all these differences. The fact that ad descriptions are small, for instance, can be mitigated by augmentation techniques similar to what anchor text does for web pages. However, all this work has limited impact. Contrary to web search, ads are not selected purely based on relevance. Most ad selection engines try to maximize revenue. Therefore, they select ads based on the product of the click probability and the value that advertisers are willing to pay for a click:

$$\text{Revenue}(A, Q) = \text{ProbabilityOfClick}(A, Q) \times \text{Bid}(A, Q)$$

For each ad selected by the ad's backend we compute the expected revenue and display the top ads according to that criteria. The quality of the ad selection, or how well an ad is related to a search query, only influences the probability of the user clicking on the ad. It is very likely that a user that searched for "latin canon" has no intention to buy a printer and will not click on printer ads. But showing relevant ads in queries that have commercial intent is very effective, and increases the probability of a click. The bid, however, is a lever that is solely in advertisers' hands. Bidding exorbitant amounts of money for ad A to appear for query Q, may place ad A at a higher position, even in the presence of other more relevant ads.

Although this revenue formula is for search advertising, all forms of online advertising will select ads based on the expected revenue that they will generate. This means that, unlike web search, ad selection can be very nondeterministic. Despite the guardrails imposed by the ad selection algorithms, advertisers have a lot of control of which ads will be shown for a given web query or for a given demographic using a social media site. With that in mind, let's step back from the algorithms and look more broadly at the role of advertising in the world.

The digital-only myth "online advertising"

We have seen that it is very difficult for website owners to manipulate the placement of their content on search results pages. Search advertising, and all other types of advertising, online or offline, is based on paid placement. Although there are guardrails imposed by the ad selection engines, such as not placing ads on non-commercial queries like "latin canon" or "page rank paper," and algorithms select ads based on their relevance, at the end of the day, advertisers are paying to display their ad campaigns at the top of the results pages. Even in offline advertising there are guardrails, as magazines won't display ads that are offensive or that do not align with their policies. And like in search advertising, in the offline world, ad selection also tries to maximize revenue. There are limited advertising spots in an issue of the Rolling Stone magazine or during the Super Bowl, and whoever pays the most will take them.

The ads we see are not selected based on their intrinsic qualities. The primary criteria is advertisers' willingness to foot the bill. There is no determinism in the selection. From one minute to the next, a completely new set of ads may be displayed for the same query issued by the same user. This volatility is due to the **low barrier of entry**. It is impossible for any website owner to make their content appear at the top of the search results page for the query "New York Times." The official New York Times website will likely always be the first displayed link. In the case of search advertising, anyone can displace this order and have their results shown at the top, if they are willing to bid high enough. The same is true for offline advertising. Anyone can run an ad during the Super Bowl, even if all publicly available reviews for their products or services are terrible.

Although this was not the case in the times of John Wanamaker, currently there are many robust mechanisms that allow us to find information about products and services. Besides web search, there are online reviews, discussion forums, and product search engines. None of those rely on paid placement. Review apps and sites are, in general, not subjected to information cascades. Most of the reviews by users reflect their direct experience with the product or service. Absent fraud, companies have no direct way to impact the ratings of their products, in the same way website owners cannot directly control the placement of their pages in search. Even social media, which is not robust for information dissemination due to information cascades, may be a good platform for companies to showcase their products.

It is important that we, as consumers and users of these technologies, understand how the platforms we use work. In many cases there are disconnects of what users expect from these platforms and how they operate. We've seen that consuming news from social media may not be a good idea, as the content is highly fragile. We learned that information cascades are easily dismantled. In social media platforms, stories are not selected by their intrinsic qualities or by how authoritative the author is, but by herding. This doesn't mean social media is bad for all scenarios. We've already seen it works well for finding jobs. It is also a viable way for companies and individuals to present their products and services to the world. Many social platforms, including Pinterest and Instagram, allow users to show their creations, promoting products in an organic way.

Many users may click on search ads and buy products thinking they are seeing search results. Only one in six users know the difference.[14] And these ads are not being displayed based on their intrinsic qualities, but based on the size of the advertisers' wallets. This means that many users may be buying products and services without all of the relevant information at their disposal. They could, instead, consult product review sites, browse Instagram and Pinterest for ideas, or search for blogs or the official company websites.

I've worked on online advertising for several years, and I understand the need companies have to promote their products and services. However, the abundance of platforms we have at our disposal, including specialized review

---

14. https://www.wired.com/2005/01/users-confuse-search-results-ads/ and a more recent article from 2021 https://informationmatters.org/2021/11/google-search-results-theyre-all-the-same-right/ (Last accessed in February 2025).

platforms, like Goodreads for books and TripAdvisor for hotels, restaurants, and tourist attractions, raises the question of what role paid advertising has in society. Advertising is the highest source of revenue for several tech companies. They are willing to give us many "free" services, such as email, maps, chat applications like WhatsApp, and web search, in exchange for data. In *Homo Deus*,[15] Yuval Noah Harari defines this as **dataism.** A new ideology in which data is the main asset, in the same way capital is the main asset in capitalism. The data these companies collect allow them to provide us with better services, but it also helps their online advertising platforms. Knowing our browsing patterns and interests helps them maximize the probability of a user clicking in an ad.

We often forget that online advertising is a myth. We can, as society, decide which platforms have a positive effect in our lives and which don't and, therefore, should be redefined. There are alternative ways to disseminate products that are both less costly and provide more information to consumers. Dollars spent in advertising can be spent to make these platforms better. Understanding how these systems work will enable us to design the solutions we want. These solutions can be both cheaper and better. It is a win-win situation. In the next chapter we'll move on to explore the complexities of computer systems in more detail by analyzing cloud computing.

---

15. *Homo Deus: A Brief History of Tomorrow*, Yuval Noah Harari, Harper, 2017.

4.

Cloud computing

Timesharing, virtual machines, polka dot sweaters, the Ford Pinto, monkey wrenches, and sustainability

"Would you make a ship sail against the wind and currents by lighting a bonfire under her deck? I have no time to listen to such nonsense."

Napoleon Bonaparte

As technology becomes a more integral part of our lives, we sometimes take for granted how several of our daily activities depend on a complex web of systems, many of which we may not even know exist. Napoleon Bonaparte's skepticism about steamships, not understanding the inner workings of engines and calling them "a bonfire" under the ship's deck, highlights that for centuries we have been relying on technology we don't fully understand. Cloud computing is one technology that has been the source of confusion for many over the recent years.

Surveys have shown that only 16% of Americans understand that cloud computing is related to networked computers that can be used to process and share information.[1] The most popular response, by 29% of

---

1. https://www.forbes.com/sites/joemckendrick/2012/08/29/most-americans-dont-understand-cloud-computing-does-it-really-matter/ (Last accessed in February 2025).

the respondents, is that it is either an actual cloud or somehow related to the weather. Although this study is from 2012, and understanding may be improving over the years, I think most still don't understand what cloud services are and how they work. However, despite this confusion, most of us are constantly interacting with cloud computing services. I'm doing so as I'm typing these very words. Whenever we take pictures, answer emails, or purchase anything online, we are using computing services hosted on the cloud. We call these **cloud services**.

What further complicates things is that most of these cloud services have dependencies on other services. We've already seen that a search results page combines results from different backends, including web search, ads, maps, and Wikipedia. Each of these backends is a distributed service in itself, running at scale. In fact, to compose a search results page we need to make hundreds of calls to different services. Most users understand that when conducting an online search that they are using Bing or Google, but they have no idea about these other internal or external services that need to be invoked. The Crowdstrike outage in July 2024 caused 8.5 million servers worldwide to be unable to restart. It was the largest technology incident to date.[2] The cause of the problem was a faulty software in Crowdstrike's antivirus, a service

---

2. https://en.wikipedia.org/wiki/2024_CrowdStrike-related_IT_outages (Last accessed in February 2025).

that most users didn't even know they had installed on their servers.

Several servers running the Windows operating system have Crowdstrike's antivirus installed. Moreover, it runs as a privileged service as part of the Windows boot process, meaning it runs during the machine start up. A problem with any of the components that run during the boot process is that they may cause the machine to crash and stay in an undesirable state. The popular term for this is "blue screen of death," or simply "blue screen," as the computer screen turns blue and shows a nasty message, sometimes accompanied by the sad face icon.[3] There was a software bug in a new version of the Crowdstrike software deployed that day that caused many servers to display the blue screen.

As Crowdstrike is primarily used by organizations, most personal computers were unaffected by this incident. Right after the bad software update happened, Windows servers in the public clouds, such as Microsoft Azure and Google Compute Engine, were impacted. Crowdstrike soon reverted the update to prevent further damage, but several machines remained inoperable and required some form of remediation. The overall impact was pretty broad. Globally, more than five thousand flights, 4.6% of the flights scheduled for that day, were cancelled. The incident affected 60% of all the Fortune 500 companies.

---

3. https://en.wikipedia.org/wiki/Blue_screen_of_death (Last accessed in February 2025).

The estimated total damage for the top five hundred U.S. companies, excluding Microsoft, reached $5.4 billion.[4]

There were several actions that could have prevented this incident. First, the software bug could have been prevented with more engineering rigor from the Crowdstrike engineering team. It is very hard to predict every possible corner case when you are writing software. Bugs are expected to happen. But with a good engineering process, validations catch many of the obvious ones that happen pervasively. This bug seems to fit that category. It was not a corner case, but a prominent bug that affected everyone and could have been prevented by testing. Second, it is common practice for software companies to release software updates in a gradual manner. Starting with a small subset of the users and checking for issues, and only proceeding to larger and larger groups of users if all health signals are good. These health signals are the computer equivalent of monitoring the patients signals during a surgery. Crowdstrike did not use this monitored crawl-walk-run strategy for this update.

This situation highlights not only how interconnected the world is, but also how complex all the software abstractions we use on a daily basis are. This complexity comes from the layered structure of software systems. There are many abstractions between the hardware and the systems we interact with. With distributed systems,

---

4. https://www.bbc.com/news/articles/ce58p0048r0o (Last accessed in February 2025).

that complexity is compounded. An examination of cloud computing in more detail will help us understand some of these layers a bit better.

It is not about white fluffy things in the sky

Early computers were large machines that occupied entire rooms. And although there were other computers before its time, the very first programmable, general purpose, digital computer was the ENIAC (Electronic Numerical Integrator and Computer). Being programmable and general purpose means that the ENIAC implemented the computational model proposed by Alan Turing. It could be programmed to compute arbitrary functions on integers, using the same building blocks that still prevail in today's modern computers.[5]

The ENIAC was located at the University of Pennsylvania. The project was completed in 1945 and cost $487 thousand, which would be equivalent to $6.9 million dollars today. It weighed 27 tons and occupied three hundred square feet. If we compare its computing power, the ENIAC could perform about five thousand instructions per second, as opposed to the 3 billion in the commonly available 3GHz processors of today. These five thousand instructions per second were much better than the previous way of doing things, which was employing human computers. These were men and women whose

---

5. https://en.wikipedia.org/wiki/ENIAC (Last accessed in February 2025).

jobs were to perform manual calculations. Human computers had widespread use, including even important missions such as the Manhattan Project.[6]

As the ENIAC and all early computers were very expensive, researchers soon started focusing on how to extract the most value from them. These early computers processed one program at a time. That meant that while the program was reading its input or printing, the processor was idle, not doing any computations. One of the first attempts to solve that was multiprogramming. Several programs were fed to the computer at the same time, with different priorities assigned to them. Every time the processor became idle, as the program was copying data or doing something else, the next program with the highest priority would take over the processor.

Universities soon started developing methods to democratize access to computing resources amongst students and faculty. **Timesharing** allowed users sitting at terminals to access the central computer, which allotted a slice of its processor time to the programs of each of the users. Timesharing dominated academic computing until personal computers became wildly available in the late 70s.[7] In his latest book, **Source Code**,[8] Bill Gates recounts that one of the business ideas he and Paul Allen

---

6. https://ahf.nuclearmuseum.org/ahf/history/human-computers-los-alamos (Last accessed in February 2025).
7. https://www.cs.cornell.edu/wya/AcademicComputing/text/earlytimesharing.html (Last accessed in February 2025).
8. *Source Code: My Beginnings,* Bill Gates, Knopf, 2025.

had before starting Microsoft was related to timesharing and democratizing access to computing resources.

With the advances in hardware since the days of the ENIAC, computers not only became much faster, but also much cheaper and smaller. The processors we carry in our pockets today are more powerful than the central computers Bill Gates and Paul Allen had to access through terminals before creating Microsoft. The advances in processing power allowed organizations to buy and maintain their own servers to support their applications and databases. Over time, with the digitalization of most operations, from HR, to supply-chain, and sales, organizations were left with the burden of taking care of an increased number of servers and applications. Additionally, the products and services offered by many organizations also became more digital.

It was one thing to operate a physical store in the nineties. Although store owners had computer systems to assist with their operations, they were largely unaffected if these systems became inoperable due to hardware failures or software bugs. It was annoying to have issues with your payroll or stock control systems, but by and large, there were alternative manual processes that worked and the store continued functioning with no impact to customers. The digital stores of today, and most modern businesses, have a major problem if they face "computer issues." In the extreme case, users cannot access the store, with huge potential revenue and customer service impact. I used to lead engineering for a payments company and experienced first hand the impact any blip on our system had on merchants and customers.

This demand for systems being "always on" places new responsibility on the engineers overseeing the company's systems. They must provide 24x7 support and guarantee the robustness of their applications. Failures on a server or on the network that connects these servers together, and cybersecurity, became critical for the businesses' success. To cope with these ever-increasing demands, organizations would have to staff sophisticated technology teams, capable of guaranteeing business continuity in the presence of all possible failures. That is when cloud computing came to the rescue.

Big tech companies that spent the late nineties and early 2000s building large scalable services, such as Amazon's online catalog and store and Google and Bing's search engines, soon realized that all the expertise they developed running worldwide distributed systems was very valuable. Moreover, it could be packaged and offered to customers. This would allow customers to focus on developing their products and services, while being freed from the work needed to maintain servers, replace them when they failed, make sure these servers were connected to the internet, and make sure they were secure. Providing computing resources to third party companies became known by the intuitive name of **cloud computing**.

By centrally managing computing resources, cloud computing companies and their customers benefit from economies of scale. Microsoft can buy servers in bulk to meet the demands of all of its customers, more cheaply than any of its customers would be able to do individually. The same is true for repairs. It is more efficient for the cloud providers to develop services that monitor hardware failures and devise operational procedures to fix them, than every organization in the world having to do that on their own. These and other efficiencies allow cloud providers to offer computing resources cheaper than individual customers would be able to do on their own, while still making a profit. A win-win situation.

Cloud computing does not solve all the problems that businesses face. Organizations must still have technology teams to develop and maintain their products, but a lot of the burden of operating online services now falls on the hands of cloud providers. Engineers don't need to worry about servers failing or getting disconnected from the internet. With their extra time, they can relax, look at the sky, and appreciate the clouds. Wait, maybe that's where the name came from.

Is it just timesharing again?

In the early days of computing, users were sharing a single processor. The ENIAC and early computers could execute only one program at a time. **Timesharing** allowed the processor to be used by many developers, but there was only one processing unit per machine. This changed drastically over the years. Now servers have multiple processing units, called **cores**. The Intel Granite Rapids server, released in 2024, has 128 cores. This means that many programs can be running simultaneously on the same server, each of them fully using its assigned core. In timesharing, programs compete for a share of the processor. In multi-core servers, we have many more processor cores available to run programs, and each core can still be timeshared.

In the early days, users submitted programs they needed to run and received the program output in return. Although the abstractions cloud computing provides now are varied, they are all very different from the early timesharing services. The most popular abstraction used today is the **virtual machine (VM)**. From the viewpoint of the cloud service customers, virtual machines are just like regular servers. The only difference is that instead of being physically under the desk, a virtual machine may be across the globe, in one of the many data centers the cloud provider may have around the world. Building and maintaining these data centers, which are buildings full of servers, and all the required equipment to make sure the servers run efficiently, such as generators and cooling systems, is one of the burdens cloud providers remove from the hands of their clients.

There is another important difference between physical servers and VMs. There can be multiple VMs per physical server, whereas physical servers cannot be easily split amongst different users. Take the case of Intel Granite Rapids with 128 cores. Cloud providers may decide to run a single VM using the entire server, or 64 VMs with two cores each, or 128 VMs with one core each, or any other combination. For instance, we could have 29 one-core VMs, 16 two-core VMs, one 16-core VM, and one 48-core VM all deployed in a single 128 core server. In this case, 125 out of the all of the server's 128 cores would have been assigned to VMs, while three cores would have been left idle. Figure 6 illustrates this scenario.

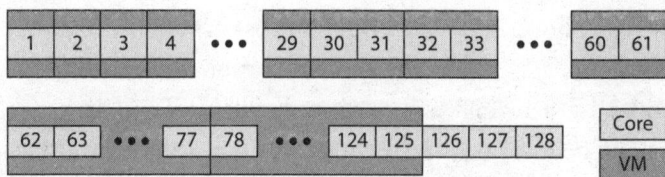

Figure 6. A possible allocation of VMs with one core, two cores, 16 cores, and 48 cores. The last three cores, numbered 126, 127, and 128, are left idle.

If we think about efficiency, this flexible mapping between VMs and servers is a bonus. With more powerful hardware, servers are increasingly growing in the number of cores, memory size, and every other dimension. Therefore, without the assistance of VM abstractions, it would be increasingly more difficult for software developers to write programs that fully utilize all the resources of a server. With VMs, software developers just need to worry about fully utilizing the VM's resources, as opposed to the full server's resources. If resources are left idle, they are wasted. Ideally, to minimize waste, all computational resources would be used all the time. Since cloud providers can pack VMs from multiple clients into the same server, they have a higher chance of fully utilizing the servers and minimizing waste. We'll come back to this point in a bit, but first let's take a deeper look at what we mean by utilization.

What happens when users fall asleep?

Server cost is a large part of the spend for any online service. For a service to be the most efficient it can be, it is imperative that the servers it runs on are well utilized. If on average a server is only 50% utilized, its cores would be idle 50% of the time. This is a simplification, as servers have other components, like memory and storage disks, and graphical processing units (GPUs). GPUs are popular today because they are used for AI and other high-performance computing (HPC) workloads, such as scientific computing and even high-frequency trading. But for clarity of presentation, let's just focus on cores for now. Assuming the 50% utilization number, we could consider buying only half of the servers we currently have and running them at full utilization. This would save us a lot of money, but would be difficult due to a few reasons.

Online services, such as Bing and Amazon, don't have consistent utilization throughout the day. Their utilization patterns are based on user behavior. Web search, for instance, has a diurnal pattern, with servers being highly utilized for half of the day, while users are searching, and having much lower utilization for the other half, when users are sleeping.[9] Search engines then need to provision enough machines to sustain their **peak load**. In the case of web search, this would be the time of the day that has the largest volume of user queries. Otherwise, they would not be able to respond to all user queries during peak. And by having provisioned this large number of machines, if they don't do anything about it, these servers would go underutilized during off-peak hours.

Another reason that achieving full utilization is hard is that interactive services, such as web search or gaming, wouldn't perform well if all the cores it consumes were highly utilized all the time. In interactive services, users act and wait for the system to respond. Any slight increase in query volume or unexpected failure would impact users by resulting in slower response times. If we go back to the Walgreens lines analogy, when the system is at 100% utilization, any delay, such as a cashier having to step out for a few minutes or a customer with an unusually packed shopping cart, would add delays to the system. Therefore, online services intentionally buy enough servers to operate at 50-80% utilization of their peak instead of at 100%.

There are computer systems that are not interactive. They perform calculations that do not require user inputs. Search engines compute new values of PageRank for every web page daily. These computations run in the

9. Towards Energy Proportionality for Large-Scale Latency-Critical Workloads, D. Lo, L. Cheng, R. Govindaraju, L. Barroso, and C. Kozyrakis. Proceedings of the 41st ACM International Symposium on Computer Architecture, 2014.

background without human intervention. Answering search queries, online shopping, and gaming, on the other hand, require user interaction, and user satisfaction declines if the systems are slow. Non-interactive services can normally run at higher utilizations than online, interactive services, as any blip in the system wouldn't directly impact users.

Without the help of cloud computing, organizations must decide how many physical servers to buy to run all their interactive and non-interactive services. This means that Bing engineers need to calculate how many servers they need to answer web search queries, at the peak hour, and how many they need for computing PageRank and anchor text, generating the inverted index, and all the other tasks they need to do in the background to keep the overall system functional. They must also consider the expected utilization of each of these services. Search backends have to run at 50-80% at peak, while PageRank can run at a higher utilization.

By investing in engineering, they can develop systems to keep the utilization of the systems high. I've personally worked on solutions that would take advantage of the diurnal patterns of search traffic and use the spare capacity of the search servers to run background computations, such as PageRank, when users are sleeping and the volume of search queries is low. However, doing this is hard and most organizations don't have the resources to focus on this level of optimization. Cloud computing helps alleviate this problem.

One advantage of using the cloud is that organizations don't need to buy servers anymore. They now run their services on VMs. And they can create new VMs or destroy existing ones based on their needs, on the spot, instead of paying for servers upfront. With cloud computing they pay as they go, and cloud providers bill per usage. This provides other options to service owners. Once you buy physical servers you are stuck with them, even at night when your users are sleeping. If a service is running on the cloud, on the other hand, it is possible to reduce the number of VMs when traffic is low. This is a lot simpler to do, than to carefully orchestrate when services should run based on their expected utilization and usage patterns.

Service owners can specify the desired features, including number of cores and memory size, of each of its VMs. They can also control the number of VMs running at any given time, which allows them to manage the utilization of their services in real time. Without the cloud, they had to be very diligent when buying new servers to make sure they have enough capacity for their services to run smoothly. The cloud allows them to revisit this as often as they need. In the same way the e-book provides added functionality over its physical counterpart, such as showing interesting passages highlighted by other readers, the same is true for VMs. With the digital transformation from physical to virtual machines, clients can now buy or discard VMs much more easily than physical servers.

Service owners no longer need to think about the utilization of physical servers, but the problem doesn't go away. It becomes the responsibility of cloud vendors. This is good news, as it lifts the burden from most of the organizations, transferring the work to a handful of tech companies, who have the engineering resources to do a good job solving the problem of utilization. In addition to concentrating the problem in the hands of a few companies, cloud providers have a better chance of achieving higher levels of resource utilization. While companies have a limited number of services to place at their servers, cloud providers have all the services from all of their clients. This is a much larger set, which enables a lot more possible mappings between VMs and physical servers. It turns out that the algorithm that places VMs on physical servers is very similar to another algorithm, used by travelers all over the world when packing their bags.

I also need an extra sweater

I like to travel light, but many people, including my daughters, love to pack as much as possible for any trip. There is always a chance you will need a sweater for your summer vacation in Jamaica. And you better take both the green and the polka dot sweaters, as you never know what your mood will be when you finally get to wear them. Just like servers have a limited number of cores and memory, bags have their own maximum capacity. There are limits on how many VMs can be placed in a server and in how many sweaters fit in a luggage. **Knapsack** is one of the classic computer science problems that nerds like me learn in CS101.

In the formal definition of the problem, we have a list of items we'd like to pack. Each item has a weight and an associated usefulness score. For instance, we may have:

- Encyclopedia: weight 25kg, usefulness 10
- Shoes: weight 2kg, usefulness 9
- Jeans: weight 3kg, usefulness 8
- Laptop: weight 4kg, usefulness 7
- Book: weight 2kg, usefulness 6
- Green sweater: weight 4kg, usefulness 5
- Polka dot sweater: weight 3kg, usefulness 4

A couple of disclaimers. The weights in this example don't reflect reality, I chose these for simplicity of the presentation. Real life shoes weigh between 250-400 grams if you are curious. Also, in my opinion, the usefulness of the sweaters is zero, but I chose 5 and 4 to keep my daughters happy.

Besides the list of items with weights and scores, the other input to the problem is the total capacity of the bag. In this example let's consider it is 10kg. The goal is to maximize the usefulness of the bag while not exceeding its capacity. We also can't split items. Either we take the polka dot sweater or we don't.

If we could take only one item, the problem would be easy. Just take the most useful item that fits on the bag. In this example, although the encyclopedia has the highest usefulness score, it doesn't fit in the 10kg luggage. We would have to go down the list to the next most useful item and take the shoes.

As we are allowed to bring more than one item, we must consider all possible combinations. Here are the two best possibilities:

- Shoes, jeans, and laptop: combined weight 9k, combined usefulness: 24
- Shoes, jeans, book and polka dot: combined weight 10k, combined usefulness: 27

The second option has the best combined usefulness score and we'd end up taking the polka dot sweater. To find the best solution we need to analyze every possible combination, which grows exponentially with the number of items. With five items there are 2^5 or 32 possible combinations. With 50 items, the number of combinations is 2^{50}, which is over one quadrillion. For any realistic application, such as mapping VMs to physical servers, 50 items to pack is a very low number, so the possible number of combinations is huge in real word scenarios.

That's why knapsack is so interesting and computer scientists are so fascinated by it. It looks quite simple but, as the number of possible combinations is exponential, there are no algorithms to solve it in bounded time. Knapsack belongs to a class of problems called **NP-hard**. NP stands for nondeterministic polynomial time. A polynomial time algorithm is one that finishes in time proportional to order $O(n^k)$, where n is the size of the input, such as the number of items to pack, and k is arbitrary constant. This is again using the order, big-O notation, we defined in Chapter 1. In the case of the bubble sort algorithm, k is 2, so we can finish sorting very quickly. The time to solve knapsack would be proportional to the number of combinations which is $O(2^n)$. This is not polynomial time, it is exponential. For typical input sizes in which n is large enough, 2^n is much greater than n^2. In practical terms it means the computer program would never finish for any realistic value of n.

The analogy for NP-hard problems is a very large maze. Since the number of combinations is too high, there is no way to find a solution to a maze in bounded time. But it is possible to design an efficient, polynomial time algorithm for a computer to check if a given solution to the maze is correct. The good news is that computer scientists already spent a lot of time thinking about NP-hard problems and came up with algorithms that find good approximate solutions in polynomial time.

One simple approximation of knapsack, proposed by American mathematician George Dantzig, is to sort the items by the ratio of usefulness divided by weight and select the items in order. In our example the order would be:

- Shoes: weight 2kg, usefulness 9, ratio 4.5
- Book: weight 2kg, usefulness 6, ratio 3
- Jeans: weight 3kg, usefulness 8, ratio 2.67
- Laptop: weight 4kg, usefulness 7, ratio 1.75

- Polka dot sweater: weight 3kg, usefulness 4, ratio 1.33
- Green sweater: weight 4kg, usefulness 5, ratio 1.25
- Encyclopedia: weight 25kg, usefulness 10, ratio 0.4

We'd start with an empty list of items in the bag. In each interaction of the algorithm, we would check if the next item in the sorted list fits in the bag, taking into account the items that have already been selected. If it does, we'd add it and adjust the bag's remaining capacity. These would be the steps:

1. List is empty. Available capacity is 10. Is $2 \leq 10$? Yes. Add the Shoes. New available capacity is 8.
2. List is {Shoes}. Available capacity is 8. Is $2 \leq 8$? Yes. Add the Book. New available capacity is 6.
3. List is {Shoes, Book}. Available capacity is 6. Is $3 \leq 6$? Yes. Add the Jeans. New available capacity is 3.
4. List is {Shoes, Book, Jeans}. Available capacity is 3. Is $4 \leq 3$? No. Don't add the Laptop. New available capacity is still 3.
5. List is {Shoes, Book, Jeans}. Available capacity is 3. Is $3 \leq 3$? Yes. Add the Polka dot sweater – yay! New available capacity is 0.
6. List is {Shoes, Book, Jeans; Polka dot sweater}. Available capacity is 0. Is $4 \leq 0$? No. Don't add the Green sweater. New available capacity is still 0.
7. List is {Shoes, Book, Jeans; Polka dot sweater}. Available capacity is 0. Is $25 \leq 0$? No. Don't add the Encyclopedia. New available capacity is still 0.

If our implementation of the algorithm was smart, it would have stopped after step five, when the bag became completely full. This simple approximation of the algorithm is efficient and produces very good results in many cases.[10] It was able to produce the optimal solution for our example, but that was luck. There are scenarios for which it would produce a good solution but not the best. Our understanding of knapsack will be useful for us to understand how cloud computing providers allocate VMs. But first we must understand the Ford Pinto case.

The Pinto leaves you with that warm feeling

Lily Gray died hours after the car crash. Her passenger, thirteen-year-old Richard Grimshaw, survived but suffered disfiguring burns. He had to undergo multiple operations to reconstruct his ear and nose, using skin from the rare parts of his body left unscarred. This was 1972 and she was driving a Ford Pinto that exploded after she was rear-ended. The Pinto's gas tank ruptured during the collision. Gasoline vapors spread through the passenger compartment and a spark ignited the tragic explosion.[11]

Ford was under pressure from Volkswagen and Japanese automakers in the growing market for

---

10. Please see a visualization of this implementation of the knapsack algorithm at https://digitalagencybook.org/visualizations/knapsack.

11. https://www.tortmuseum.org/ford-pinto (This is a very interesting article. Most of the content about the Pinto case came from here. Last accessed in March 2025).

subcompact cars. It needed a model to compete with the Volkswagen Beetle. Ford's board approved the Pinto's plan in 1969. Lee Iacocca, Ford's president at the time, wanted the Pinto to be priced at less than two thousand dollars, which is roughly $17 thousand in current value. The other main constraints were the weight, which needed to be under two thousand pounds, and time to market.

The desire to release the car in 1971 imposed very strict design and production schedules. The Pinto's development cycle was only 25 months, nearly half the industry's average of 43 months. This was the shortest schedule ever executed by any automaker at the time. Several steps normally done sequentially, such as design and tooling, were done in parallel. And there was very little time in the schedule reserved for safety testing. Lawsuits confirmed that Ford's rushed timelines contributed to the car's safety issues. A review of internal tests performed before the car reached the market showed that every collision over 25mph caused gas tank ruptures, often igniting fires.

The Grimshaw and Gray families filed a tort action against Ford. In addition to winning their claims for injuries, Ford also had to pay punitive damages for endangering the lives of thousands of Pinto owners. Grimshaw v. Ford is just one of several lawsuits filed by survivors and their families. It is estimated that between five and nine hundred people died in similar accidents. These deaths could have been prevented if the gas tank design flaws were fixed during safety testing.

A damning investigation by the National Highway Traffic Safety Administration (NHTSA) in the late 1970s led Ford to recall over 1.5 million Ford Pintos produced between 1971 and 1976, along with approximately 30,000 Mercury Bobcats. In the same year, General Motors also had to recall Chevettes for similar fuel tank issues. The Pinto case became emblematic of automotive safety failures. It is tragically ironic that one of the radio advertisements for the Pinto included the line, "Pinto leaves you with that warm feeling." This sad story underscores the dangers of prioritizing cost, weight, and time to market over safety. Another key lesson is that investing in thorough testing and validation is far more efficient, and ethical, than allowing customers to become unwitting testers. In Pinto's case, inadequate safety testing resulted in a fragile product, and some customers paid with their lives. The recent Crowdstrike outage serves as a modern reminder of the consequences of inadequate testing in a different domain.

Although the Crowdstrike outage was much less tragic than the Ford Pinto accidents, it was still a very costly way to find out about a problem. The investment in increased engineering rigor, including better testing and gradual software updates, would have been much smaller than the financial impact of the incident. This type of tradeoff happens not only with testing. It is pervasive in engineering. The investment Amazon made in its supply chain, is another example. A central investment by Amazon lowered the shipping cost for all of its clients.

I like to call this a **platform investment**.[12] By building a central platform, be it for software testing, supply chain, or car safety, we can improve quality and reduce costs.

As we'll see next, the problem of mapping virtual machines into servers is an optimization problem with many, often conflicting, facets. The platform investments made by cloud providers lowered the barrier of entry by solving hard engineering problems that were in the hands of their clients.

Virtual machines run in real machines

From the user viewpoint, a VM behaves just like a physical server. You can use it to run your favorite software, reboot it, or anything else you could do with a physical server, apart from kicking it when you get frustrated. Kicking is impossible because the VM is "running" in a server located in one of the cloud provider's data centers. You can think of the VM as a software service that gives you a slice of a physical server.

As we've seen before, VMs come in different sizes, such as:

- size-1: 1 core, 2GB of memory, 30GB of storage
- size-2: 2 cores, 2GB of memory, 40GB of storage
- size-3: 4 cores, 4GB of memory, 80GB of storage
- size-4: 8 cores, 16GB of memory, 200GB of storage

---

12. *A Platform Mindset: My lessons from developer to CTO*, Marcus Fontoura, 8080 Books, 2025.

Cloud providers normally offer a vast selection of VMs, with different sizes and configurations, so that their clients can select the best options for running their services. If a given client runs a VM of type size-2 in the example, it would have 2 cores, 2GB of memory, and 40GB of storage fully dedicated for running its applications. Even if there were other VMs, possibly from different clients, running in the same machine at the same time, these VMs wouldn't be able to access the resources assigned to the VM in question.

If cloud providers had only one server in which VMs could run, the problem of allocating VMs into servers would be very similar to knapsack. As long as there are resources available we'd assign new VMs to run on the lonely server. When we run out of capacity, we have to make rational choices. Let's consider that this server had a total of ten cores, 20GB of memory, and 300GB of storage. If we wanted to run two size-1, one size-2, and one size-4 VMs, we'd have to select one of the three options shown in Table 1, as the server wouldn't have enough resources to run the four VMs simultaneously.

VMs allocated	Cores used	Memory used	Storage used
option 1: size-1, size-1, size-2	4	6GB	100GB
option 2: size-1, size-1, size-4	10	20GB	260GB
option 3: size-2, size-4	10	18GB	240GB

Table 1. Three possible allocations for the four requested VMs.

When the sum of the resources allocated to VMs is smaller than the server's physical resources, we have wasted resources. Each of the three options we have leaves us with some wasted resources. The option in the first row uses only four out of the ten available cores, and leaves 14GB of memory and 200GB of disk unused. The option in the second row consumes all cores and memory, leaving 40GB of storage unused. The last option uses all the cores, but wastes 2GB of memory and 60GB of storage space.

As the servers are the most expensive expenditure cloud providers have, the algorithms used to map VMs to physical machines try to minimize the amount of resources that go unused to keep costs down. They achieve that by defining a metric of waste that takes into account the scarceness of each of the different resources. If cores are the most precious resource, followed by memory and then storage, we could assign weights 6, 4, and 1 to these different resources, resulting in formula:

$$\text{Waste} = 6 \times \%\text{unused cores} + 4 \times \%\text{unused memory} + \%\text{unused storage}$$

With this formal definition of waste, we could allocate VMs into servers using a variation of knapsack that, instead of trying to maximize the usefulness score, would try to minimize waste. If we consider our example, waste for the three different scenarios would be:

Waste for option 1: $6 \times 0.6 + 4 \times 0.7 + 0.67 = 7.07$
Waste for option 2: $6 \times 0 + 4 \times 0 + 0.13 = 0.13$
Waste for option 3: $6 \times 0 + 4 \times 0.1 + 0.2 = 0.6$

Option 2 would be the one that minimizes waste the most, with option 3 being a close second. Option 1 would be the most wasteful as it leaves 60% of the cores, which is the most scarce resource, unused. Cloud providers have so many servers that even a small reduction in waste can represent millions of dollars in savings. This brings us back to reality. The problem cloud providers face is a bit more complex than this simple example.

They have a lot more than a single server and must minimize waste across all of them. As these servers are spread across many data centers around the world, clients are able to select the geographical location where their VMs should run. This is important both for response time, as some services may have strict performance guarantees, and for compliance, as some industries may require user data to be stored in data centers within the country or region. In addition to selecting the geography, service owners may also specify more strict **placement constraints**. These placement constraints restrict the set servers that are eligible to run a VM.

One scenario for which placement constraints are very useful are distributed systems that must manage the resiliency of their services. As we've seen, services like web search need more than one server. If they run in the cloud, they would need more than one VM. In the case of web search, each of the VMs would store a slice of the inverted index, which we called a **shard**. Since it would be unhelpful for Bing or Google to not return amazon. com as a result for query "amazon," they normally try to make sure that the index shard containing amazon.com is replicated in more than one VM. In the absence of placement constraints, it would have been possible for VMs containing two replicas of the same index shard to land on the same physical server. Faulty hardware in that server could then affect both replicas simultaneously.

By applying placement constraints, service owners can make sure their VMs are spread in different servers, minimizing the chances that a single failure impacts VMs that should not fail simultaneously. We all understand that servers are susceptible to failures, but there are other components in the data center that can also fail, like electrical components that keep servers energized, and the cooling system, which control the processor temperature, avoiding meltdowns during intense work. Failures in these components are more widespread, and may affect multiple servers. Sophisticated clouds may provide fine-grained controls to service owners, allowing them to place their VMs in a way that minimizes the chances of these correlated failures. Servers that fail together due to faulty data center equipment, are said to be in the same **fault domain**.

Fault domains are hierarchical. A single server is a fault domain, as it can fail independent of other servers due to faulty hardware or software bugs. Servers that operate under common data center electrical and cooling systems are in the same fault domain. And the whole data center itself is also a fault domain. A natural disaster like an earthquake could cause the whole data center to fail. In scenarios where resiliency is paramount, sophisticated placement constraints can be put in place to make sure VMs are properly balanced across faulty domains. Although placement constraints provide the ability for service owners to manage the resiliency of their system, they come with a cost.

Let's consider a data center with three fault domains. Each having ten servers. A service owner could request to deploy nine virtual machines, spread evenly, in different servers across the three fault domains. That could have been specified as:

Fault domain 1: 3 VMs of size-2 in 3 distinct servers out of the 10 available ones.
Fault domain 2: 3 VMs of size-2 in 3 distinct servers out of the 10 available ones.
Fault domain 3: 3 VMs of size-2 in 3 distinct servers out of the 10 available ones.

It may turn out that the ten servers in fault domain two are full enough to not be able to add a size-2 VM. In fact, even if eight out of the ten available servers were full enough, we'd not be able to place the three requested VMs on different servers. This example highlights one of the many tradeoffs that cloud providers face when assigning VMs to servers. The compromise between minimizing wasting server resources and maximising available capacity for future requests across fault domains.

Figure 7 shows an example where there are two servers that are able to host a new VM. The first option minimizes waste, consuming fully all the resources available in the server. The second option, despite leaving some resources unused in both servers, would leave room for a possible new deployment that requested VMs deployed across the two servers. There is no right or wrong answer here. If, instead of an allocation that needs VMs spread across two servers, we had a request for a large VM, option one may have been a better choice. Independent of which option we choose, there would be future VM allocation requests that would fail while others would succeed. Cloud providers need to understand their usage patterns and choose the option that maximizes the chances of future allocation requests succeeding.

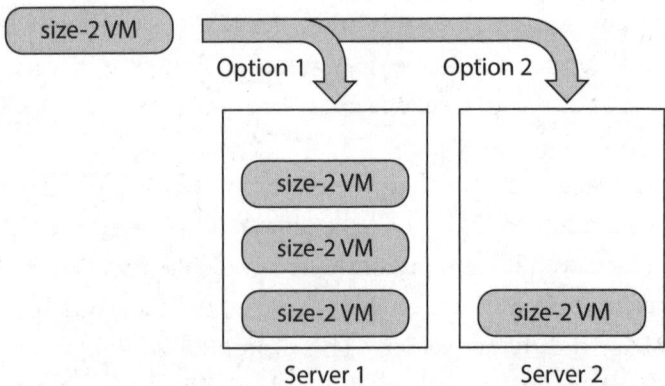

Figure 7. Tradeoff between two possible allocation options.

There is also a tradeoff that cloud clients face. Although placement constraints enable them to manage the resiliency of their services, the more constraints they place, the higher the risk that their allocation request will fail. It is better not to impose constraints and give more choices to the cloud provider allocation service whenever possible. Placement constraints are powerful, but with great power comes great responsibility. If not properly used, placement constraints will lead to unnecessary cost and inefficiencies.

Cloud providers strive to allocate all their physical resources to VMs, minimizing waste. However, they have no control of how well their clients utilize their VMs. If a client requests a very large VM that consumes all the cores, memory, and storage of a server, but leaves the VM idle, the actual utilization of the server will be low. It is in the customers' best interest to maximize the use of their VMs since they are paying for it. In the same way, it is in our best interest for us to make good use of the gym membership we pay for monthly.

Some clients are able to do a good job while others are not. The cloud provider can measure the actual utilization of the servers and may act in different ways in case a client is not fully utilizing their VMs. The easiest thing is to follow the lead of gym chains around the world – pocket the money and do nothing. The VMs are already paid for, so cloud providers could just profit without taking additional action. This is not ideal if we consider that servers are not only expensive, they also consume energy. As we'll discuss soon, minimizing the carbon footprint of cloud services is not only a noble goal, it is so important that most public cloud providers have made public pledges to become carbon neutral or even carbon negative.

A nicer option is to recommend that their clients reduce either the number of VMs they requested, or their VM sizes, or both, to match the actual resources they need, minimizing waste. This approach is called **rightsizing**. All major cloud providers offer rightsizing solutions, which analyze utilization patterns and proposed recommendations. If clients adopt these rightsizing recommendations, they lower their cloud spend and increase the average utilization of their VMs. They pay less and less physical resources are wasted. A win-win situation. Another way cloud providers can help to increase the utilization of servers is by oversubscribing resources.

Early computers introduced timesharing to allow the processors to be better utilized. Whenever a program was not utilizing the processor, a new program would take over. Modern machines have far more cores, but timesharing is still possible. Nothing prevents cloud providers from taking advantage of this and mapping more than one VM to the same core. For instance, we could have 64 4-core VMs in a server with 128 cores. In this case we'd say that the machine is **oversubscribed**. When more than one VM is assigned to a core, the programs running in these VMs will be sharing the core, similarly to early timesharing solutions. This has the advantage of increasing server utilization. The downside is the potential negative impact on the performance of these programs, should they become active simultaneously.

Cloud providers may offer regular VMs and oversubscribed VMs as different options. Service owners, interested in controlling their cost while keeping the utilization of their services high, may run their services in a mix of regular and oversubscribed VMs. A search engine may use regular VMs for their interactive search service, while opting for oversubscribed VMs for non-interactive workloads, such as the services that compute PageRank. The oversubscribed VMs are less expensive, but don't offer performance guarantees. They run in cores that are shared, which may incur delays in the computations running in these VMs. This may be fine, as a lot of non-interactive computations are not time bound. Programs running in an oversubscribed VM behave like two friends sharing a video game console for a single player game. Only one can play at a time.

Another mechanism that is available for cloud services is **autoscaling**. It automatically adjusts the number of VMs based on load. In the same way we can program a thermostat to keep the temperature comfortable, engaging the heater if it is cold or the airconditioning if it is hot, autoscaling increases or decreases the number of VMs to match the user demands. During the day, when active users and traffic increases, it could gradually increase the number of VMs dedicated to answer requests. After having allocated enough VMs to handle peak traffic, the number of VMs could be gradually reduced, matching the services's diurnal pattern.

Multiple VM sizes, placement constraints, rightsizing, oversubscription, autoscaling are different mechanisms the cloud providers offer, allowing clients to have control over the cost and resiliency of their services. By leveraging the platform investment made by cloud providers, engineers can then focus on their applications requirements. The challenging problems of keeping servers running in a healthy state and minimizing resource waste become the responsibility of the cloud platforms. Another major issue cloud providers must address is minimizing their carbon footprint. By the end of the decade, we may need 8% of the total energy in the world to power cloud services.[13] But before talking about energy, let's dive deeper into resiliency.

Throwing a monkey wrench at your service

I had the frustrating experience trying to play songs through Spotify while using Apple's CarPlay interface, in a rental car during my last business trip. Something was not working. I could see my playlist but I could not play any songs. I finally gave up and reverted back to good old FM radio. I then decided to count the number of services that were involved in that interaction. We have at least the following list of services. Don't worry if you

---

13. https://spectrum.ieee.org/cloud-computings-coming-energy-crisis (Last accessed in March 2025).

don't know some of the terms below, as this is just an illustration:

1. Car's entertainment system. My rental car had a system that allowed me to connect my phone and was compatible with Apple CarPlay.
2. Apple CarPlay software. Apple's system that connects the iPhone with the car's system.
3. The iPhone operating system. The iOS system that provides connectivity with the car, using Bluetooth.
4. Bluetooth. Service running in iOS that links the iPhone to the car's system.
5. Cellular network. Spotify uses this to stream music. It requires mobile data via your carrier, such as 4G or 5G plan.
6. Apple CarPlay application programming interface (API). It provides the mechanism for Spotify to connect with Apple CarPlay.
7. Spotify app. This is the app that was installed on my phone and I was using to play music.
8. Spotify's streaming service. All the backend services that Spotify provides, including authentication, content delivery, and recommendations.
9. Car's audio system, including the speakers. The car entertainment system must connect with it to play anything.

Each of these services may depend on other services that I don't know about. And each of them may run on a cloud provider and leverage some of the services we just talked about, like autoscaling. So, despite being frustrated that I couldn't play my music, at least I know that there is a lot that could go wrong. In my case, I think it was either a problem in the connection between Apple CarPlay and the car's entertainment system or a bug in Spotify's integration with Apple CarPlay. Who knows. Building resilient software is extremely hard. And building a system that never fails is not only hard, it is impossible.

There is not a single large scale system in the world that has never failed. However, with enough engineering rigor it is possible to minimize outages and to reduce the time to mitigate them once they occur. If the team has the mindset that incidents are expected, they will focus on the right engineering toolset to help contain and solve problems once they arise.

By embracing the philosophy that incidents are inevitable and with a focus on preparation, Netflix decided to deliberately throw monkey wrenches at their production services. Essentially, they build software systems to inject small problems into their production systems, which enables them to build automated solutions to handle these problems and be more prepared when serious incidents occur. Chaos Monkey is a system built by Netflix that randomly kills virtual machines.[14]

---

14. https://www.computer.org/publications/tech-news/research/netflix-chaos-engineering (Last accessed in March 2025).

As it was migrating from its own data centers to the cloud, Netflix had problems with the stability of some of its production services. The sudden termination of VMs, while rare, caused problems for some of the services, potentially impacting end users. Their solution was to make the unexpected termination of a VM more of a norm. Chaos Monkey works by choosing VMs at random to shoot down. Of course, this is done at a small and controlled rate so that these small incidents are not a major catastrophe. Additionally, Chaos Monkey works only during regular business hours, not to induce problems when engineers are having ice cream with their families.

Engineers don't like to solve the same problem over and over, and these small incidents help them build creative solutions that make their services more reliable. One of these possible solutions may be to use autoscaling to make sure there are enough VMs to sustain user traffic, even in the presence of a small number of VM failures. Besides Chaos Monkey, Netflix also developed other systems, including Chaos Kong, which simulates the outage of an entire data center region, and Chaos Gorilla, which simulates failures of smaller fault domains. It can be a hard sell for a tech leader to convince the company leadership that investing engineering time to cause havoc in your own systems is a good idea.

We can estimate the return on investment (ROI) of chaos engineering using a simplified financial model. Let's consider the following variables:

- Original cost: The cost of incidents if we don't invest in chaos engineering. As an example, let's imagine that in the last six months before adopting chaos engineering we had eight incidents of medium severity and two major incidents, amounting to a loss of one million dollars in damages.
- Investment in chaos engineering. The engineering cost to develop systems like Chaos Monkey and Chaos Gorilla. Let's consider it took four engineers for a few months, amounting for a total investment of $400 thousand.
- Chaos budget: the financial loss caused by the small self induced incidents. Let's say this was $20 thousand in the six months after chaos engineering was introduced. We can always put a cap on this budget.
- New incident cost: in the six months after introducing chaos engineering we had two medium severity incidents and one major incident. However, they were mitigated more quickly than before and the total cost was only $200 thousand.

The ROI would be:

Return = Original cost - Investment in engineering
        - Chaos budget - New cost
      = $1 million - $400 thousand - $20
        thousand - $200 thousand
      = $380 thousand

$$\text{Investment} = \text{Investment in engineering} +$$
$$\text{Chaos budget}$$
$$= \$420 \text{ thousand}$$

$$\text{ROI} = \text{Return} / \text{Investment} = 0.9$$

This ROI measure is negative if the investment was greater than expected return. It's zero if the investment exactly matched the return. And it is positive if the investment paid itself. In this fictitious example, as ROI is close to one, the company had almost a dollar of return for each dollar invested. This ROI calculation proposed by Netflix is a good tool to show that chaos engineering is a good investment. While this is a simplified model and doesn't account for long-term benefits like improved team readiness, better architecture, or reputational protection, it still illustrates that chaos engineering can pay off significantly. Even so, it can be a hard sell, especially when it involves deliberately triggering failures by shooting down your own VMs. But metrics like these provide a compelling argument that proactive resilience engineering is a sound investment.

Although it is impossible to eliminate incidents completely, with the right engineering rigor it is possible to reduce their frequency and to minimize their cost and impact when they happen. Relying on the platform investments from cloud providers is a good idea. One of these investments that we still haven't discussed is energy efficiency. In addition to keeping servers well-utilized, efficiently managing power is also key for running data centers in a cost effective way. And the good news is that doing a good job in power management is not only good for efficiency, it also has a positive impact on sustainability.

The physically-backed system "power management"

Besides the servers themselves, energy-related costs are the dominating factors in the total cost of ownership for cloud computing. These costs include the monthly electricity bill to keep the servers running, but also the costs of the cooling infrastructure, which prevent the servers from melting, and the power delivery infrastructure, which provides uninterrupted and reliable electrical power to all the data center IT equipment, such as the servers and the networking gear.

Power comes from the local grid, often using multiple feeds for redundancy. In addition to the utility power supply, the electrical infrastructure includes backup generators, transformers, and several other components to guarantee the servers are powered continuously, even in the presence of component failures. Power usage effectiveness, or **PUE**, measures how efficient a given data center design is in terms of energy consumption. It is defined as:

$$PUE = \text{Total facility power} / \text{Total IT power}$$

A hypothetical PUE equal to one means that all the power consumed from the grid is used for the IT equipment. In that case, there would be no overhead and all the power is used by IT operations. PUEs higher than one mean that part of the power is used for the other data center components, such as the cooling systems. Modern data centers strive to have PUEs very close to one. The lower the PUE, the more sustainable the data center design is, because less energy is wasted. Since we must have other equipment besides the IT equipment, we cannot have a PUE of one or lower.

To some approximation, both the cooling and the power infrastructure are provisioned based on the average energy that the servers will consume. Therefore, the less energy servers consume, the less infrastructure needs to be provisioned. If we assume for simplicity a PUE of one, in a data center provisioned for 32 mega watts (MW), all this power would be used for the IT equipment. A typical cloud provider data center varies between 20 and 50MWs, with some reaching 100MWs, which is a typical power consumption for a small city. For building large AI models, data centers may go way past that, getting close to 500MWs. The more power each server consumes, the smaller the number of servers we can install in a data center. To be able to compute how many servers can be installed in a 32MW data center, we need to understand how server utilization is related to its power consumption.

As we have seen, it is hard to keep servers running at high utilization all the time. It is undesirable for interactive services to run at peak utilization, since any traffic fluctuation or internal disruption, such as hardware faults, could trip the service over the edge and negatively impact users. Despite all efforts for keeping utilization high, studies show that, on average, it fluctuates between 10-50% of their maximum utilization levels.[15] The use of some of the techniques we've seen in this chapter, like rightsizing, oversubscription, and packing VMs in servers to minimize waste are useful to increase server utilization.

Servers are never fully idle. There will always be someone that wakes up in the middle of the night to search for chocolate chip cookie recipes. And, despite our best efforts, servers are never fully utilized either. But you may be asking yourself, why do we care that servers are not fully utilized? We could just provision the power and cooling infrastructure assuming, let's say, 30% utilization on average and not worry about it. The issue is that servers are not very **energy proportional**.

In energy proportional systems, consumption is correlated to utilization. They would, ideally, consume no power when idle, consume little power when activity is low, and gradually increase consumption as they get more active. These machines would have a wide power consumption range. This is not common in today's IT equipment, although it has been improving in recent years,

---

15. The Case for Energy-Proportional Computing, L. Barroso and U. Hölzle. IEEE Computer, Vol. 40, No. 12, December 2007. Many concepts in this section come from this paper.

since the issue of energy proportionally has been raised by two Google engineers, the late Luis Andre Barroso and Urs Hölzle.[15]

In other domains, there are plenty of examples of energy proportional systems. Starting with us, humans. We have an average daily consumption of about 120W. However, when resting, we consume as little as 70W. For a few minutes, we can also sustain periods of intense activity, consuming over one thousand watts. Elite athletes are able to approach two thousand watts. Another example of an energy proportional system is electric cars. They don't consume any power while idle, and consumption is roughly proportional to distance traveled, with variations based on speed and driving habits. I'm told I break too much, which is raising my personal carbon footprint.

But unlike humans and electric cars, servers are not very proportional. Even energy efficient servers consume more than 50% of their peak provisioned power at low utilization levels of around 10%. Servers are most energy efficient at their peak utilization and efficiency drops quickly as utilization decreases. This means that, if we want to be energy efficient, we should double down on the cloud platforms' efforts to increase average server utilization. It would be much better to use the 32MW available in a data center to run 32 thousand servers at peak capacity, each consuming one thousand watts, than to run 64 thousand servers at 500 watts each. At 50% power consumption, the servers would be at 10% utilization, as shown in Figure 8. We'd be wasting 90% of our computational power. This number looks bad enough, but let's take a closer look.

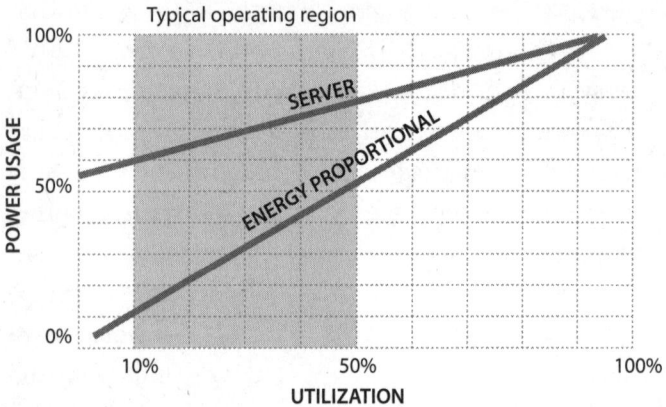

Figure 8. This figure is adapted from *The Case for Energy-Proportional Computing*.[15] It compares a typical, energy-efficient server with a hypothetical energy proportional machine. The server consumes about half of its power even when doing very little work. At high utilization, it starts behaving more efficiently in its power consumption.

By provisioning servers for 50% power consumption, our actual computational capacity would be only 10% of what we'd have by running half of these servers at full utilization. To provide the same computational capacity, we'd need to build five data centers instead of a single one. The total server costs would be tenfold and energy efficiency and data center costs would be five times worse. To be the most efficient we can, we should use our most expensive resources, which are the servers, to the fullest.

Turns out it is also the best for sustainability. If cloud providers want to be sustainable, they should strive to use every watt of provisioned power to run servers at high utilization levels. In the ideal scenario, PUE is close to one, meaning that all the power goes to the servers and IT equipment, and every server is busy all the time. Running servers at high utilization levels also means that more services can run in each data center, and cloud providers can serve more clients with fewer data centers. The most sustainable data centers are those we don't need to build. Building more data centers doesn't only require more power. All the concrete that goes into the buildings also contributes to their carbon footprint. There is work on building data centers with green concrete, but the uber point is, if possible, use your data centers to the fullest to avoid building new ones.

The part that cloud providers don't control is what types of services are running in the VMs they provide. Even if the server utilization is high, wasting computational resources to browse the deep web, spend hours on social media, or mine cryptocurrency may be a high carbon tax to pay. Ideally we would use our computational resources wisely, in services built to advance humankind. I'd love to see most of our computational resources dedicated to noble causes, such as drug discovery and investments in basic science. At the end of the previous chapters we described the digital-only myths of social media and online advertising. Power efficiency is not a myth. It is physics. Being mindful of how we spend our energy resources is imperative. Given our increased reliance on computer systems, the rise of computer energy consumption is of growing concern. Power management systems can alleviate these concerns.

In this chapter we covered a lot of deeply technical topics, given the importance of cloud computing and sustainability in our modern world. Some of the tradeoffs we explored, such as between resource waste and resiliency, are good examples of how to approach problems with a mindset of efficiency. In the next chapter we'll switch gears and talk about organizational efficiency. To design efficient systems it is imperative to design efficient organizations. One cannot exist without the other. We'll cover how to organize companies and teams to be able to deliver important projects, such as self-driving cars, web search, and cloud computing, with speed, agility, and at a low cost.

5.

Organizational efficiency

Promoting people at random, a new PUE, innovator's dilemma, and AI

> "Every new member in a hierarchical organization climbs the hierarchy until he/she reaches his/her level of maximum incompetence"
>
> Laurence J. Peter, The Peter Principle

One of my favorite quotes about organizational efficiency is attributed to Margaret Mead. It reads: "never doubt that a small group of thoughtful, committed citizens can change the world. Indeed, it's the only thing that ever has." I believe that collaboration between motivated and competent people is the only way to deliver value and positively impact the world. Organizations can be a catalyst by enabling people and projects to flourish. When not well managed, however, they can be a burden and slow down progress.

The Peter Principle,[1] proposed by late Canadian educator Laurence Peter, describes one of the many ways that organizational bureaucracy can reduce its efficiency. The principle proposes that by promoting

---

1. *The Peter Principle: Why Things Always Go Wrong,* L. Peter and R. Hull, Bantam book, 1976.

people by competence, measured by their success in previous roles, eventually put them in a position where they are incompetent. The main reason being that skills and job expectations may vary widely from one role to the other as one moves vertically within a company. Although it was intended to be a satire, the principle became popular, viewed by many as a serious investigation on how people are promoted in hierarchical organizations.

Researchers Alessandro Pluchino, Andrea Rapisarda, and Cesare Garofalo, were awarded the Ig Nobel Prize in 2010 for their computer simulations of The Peter Principle.[2] The Ig Nobel Prize is a satirical award to "honor achievements that first make people laugh, and then make them think." In their simulations they showed that, if we assume the principle to be true, it is better to promote people at random than to promote by competence. The study also showed that if the principle was not valid and the skills at one level were transferable to the next level, promotion by competence does better than random promotions. As a person that has to sit in several hours of promotion discussions every year, I find this result is a bit "sad," as random promotions would be much easier to implement and demand a lot less from management, especially in large organizations.

---

2. The Peter Principle Revisited: A Computational Study, A. Pluchino, Alessandro, A. Rapisarda, and C. Garofalo, Physica A. 389 (3): 467–472, 2010.

I believe this principle got so much attention because it is true in many situations. The skills required to be a software developer vary from what is needed to be a successful software development manager, and from each of those levels to the skills needed to be the company's CEO. Brazilian mathematician, Artur Avila, one of the winners of the 2014 Fields Medal, which is the highest honors in the field and considered the Nobel Prize of mathematics, said in an interview that in his profession the job remains the same independent of your seniority. Young mathematicians and Fields Medal winners alike do the same thing. They think about math problems.

My first job was at IBM Research, which followed the model described by Artur Avila. There were no promotions for researchers. Newcomers, like me at the time, and old timers had the same job responsibilities. All of our jobs, regardless of seniority, were thinking about important problems that could impact IBM and the overall industry, and solve those problems. Although this approach may work well for academics and industrial researchers, it is not the norm and it likely wouldn't scale well for other industries, roles, and organizations.

Management roles seek to ensure that their organization is delivering value in a cost effective way. One of the most important tasks management must do is to provide employees with the conditions to produce value without too much friction. If all the work could be done by a single or a handful of employees, it would be easy. The scale imposes taxes. As I mentioned in my previous book, *A*

Platform Mindset, with large enough groups, the organization becomes a reflection of society. Some people will not be thoughtful. Some will not be committed. Others may even be spies planning cyberattacks. Collaboration across teams becomes hard. Management's goal is to reduce these taxes with a goal similar to Margaret Mead's quote: to make large organizations feel like small groups, where smart people, motivated by the company goals, can change the world. Let's start by defining some metrics for organizational effectiveness. Although throughout this chapter I'll present examples from my work in technology companies, the concepts we'll cover are broadly applicable and not limited to tech.

A new definition of PUE

In the last chapter we learned about power usage effectiveness (PUE), which measures the fraction of the power going to the IT equipment, as opposed to being "wasted" in other parts of the data center, such as the security cameras, cooling infrastructure, and generators. Similarly, we can think of a **personal usage effectiveness**, a rebranded version of PUE, applied to team performance. The concept is very similar:

$$\text{PUE} = \text{Total work} / \text{Total value-producing work}$$

Total work is the sum of everything done in the organization by every employee. **Value-producing work** (VPW), sometimes referred to as valued-added work, was aptly named to mean everything that generates value to the company. This concept comes from **lean manufacturing** and was introduced by Toyota, as part of its Toyota Production System (TPS). It has since been widely adopted across industries far beyond automotive.[3] Lean manufacturing is a way of organizing production to maximize efficiency and minimize waste. It focuses only on activities that add value for the customer, like making the product better, while reducing or eliminating everything else, such as unnecessary steps or excess inventory. In simple terms, it's about enabling businesses to deliver high-quality products more quickly and at lower cost.

I'm a huge believer that software development is very different from manufacturing. In *A Platform Mindset*, I argue that there are **10x developers**, which are those who can generate a disproportionate positive impact on the team. Software engineering is also a creative activity, unlike an assembly line. As I mentioned before, innovation is inherently inefficient as we need room to explore new ideas and experiment. The same is true for all creative work. I had to write, and rewrite many parts of this book multiple times. All of these points make the use of lean manufacturing concepts less applicable to software, and knowledge work in general. But, despite these key differences, the PUE and VPW metrics will be useful for us to discuss productivity.

---

3. *Toyota Production System: Beyond Large-Scale Production,* Taiich Ohnoi, Productivity Press, 1988.

In software development, productive work could be defining the product, writing code, and writing tests. Of course, there are pieces of code that are more impactful than others and some engineers may be busy writing code for projects that turn out to be irrelevant. In the same way, a data center design can be very efficient and have a PUE close to 1, but all the servers are busy running irrelevant services, mining bitcoin, or scrolling through social media. For the sake of this metric, every line of code written, useful or not, will count as VPW. Employee performance reviews and career mentoring meetings, although very important to the success of the company, do not generate direct value and would not be considered as value-producing. This is akin to the cooling system. It is crucial to keep servers from melting, a necessary overhead we need to pay to be able to power the servers. In a company, there may be entire support organizations that perform non-VPW, such as HR and accounting in the case of a software company. These organizations are very valuable, but the work they do is not directly producing value, as we defined earlier. There is nothing wrong with being on the non-value producing side of PUE. HR staff and cooling systems are very "cool" (pun intended). For a very interesting and insightful discussion of work that is not useful and not value-producting I highly recommend David Graeber's *Bullshit Jobs*,[4] which discusses

---

4. *Bullshit Jobs: A Theory*, David Graeber, Simon & Schuster, 2018.

why so many people work on jobs they themselves consider irrelevant.

There are several factors we must cover to understand this simple PUE formula. Let's start with the organizational structure. Most companies and teams are hierarchical. The leader on the top has a set of direct reports, which in turn are the leaders for their subteams, and so on. At the bottom of the tree are the **individual contributors** (ICs), as shown in Figure 9. Although there are some ICs in the intermediate layers of the tree, in general, most of them will be in the lowest layer. We can assume that most of the work done by ICs is value-producing. In software engineering, there are senior ICs that will mentor junior ICs, give talks at conferences, and complain about the cafeteria food. But by and large, they are focused on writing, reviewing, and testing code, which are all VPW.

The opposite is also true, most of the work done by managers is not value-producing. Managers, of course, play a central role in the organization. In most cases, the CEO and all the C-suite are all managers. Although they may provide valuable input that will change the direction of a product, in most cases, they are thinking about strategic directions for the company and managing the team. If managers do a great job, they will hire the best people in the industry, remove any friction their employees may have, and provide a great environment in which everyone is productive. Most of these activities are not VPW.

In Figure 9, I used a value per employee to indicate their percentage of VPW. The figure shows the common pattern that, in general, this number increases as we go down in the hierarchy. Managers at the top have larger teams and, therefore, more managerial overhead. First line managers, at the bottom, may be able to juggle technical work with managing their small teams. There are two important metrics that impact team size. The number of **layers** in the hierarchy and the **span of control (SoC)**, which is how many direct employees a manager has. In this example, we have four layers and let's assume a SoC of ten, for a total of 1,111 employees, 111 of those being managers. This is a simplified and fictional example. In real life, SoC is not super uniform across an organization.

SoC = number of direct reports to a manager

| | | | | | | | Manager |
| | | | | | | | IC |

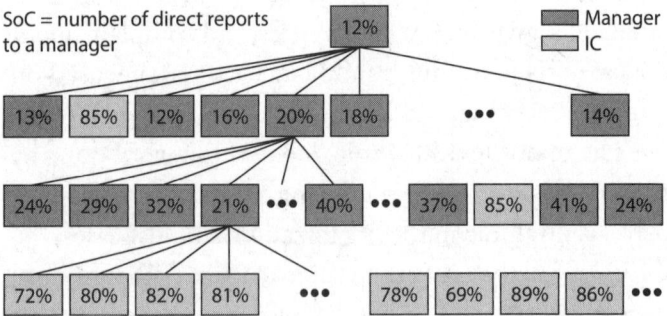

Figure 9. Example of hierarchical organization with four layers. The numbers inside the boxes represent the percentages VPW per employee. VPW increases as we go down the hierarchy, with ICs, in general, having much higher values than managers.

Calculating precise VPW percentages as shown in the figure is hard. For some professions this can be done by automation, like deriving information from development tools for software engineers, but approximations are generally good enough. As I mentioned, software engineering is a creative activity and therefore cannot be easily measured. In this example, let's assume the average VPW per manager is 30%, across all levels, and the average per IC is 75%. We can then compute the organization's PUE:

$$PUE = 1111 \times 100 / (111 \times 30\% + 1000 \times 75\%) = 1.42$$

If we had different SoC and layers, for an organization of the same number of ICs, the PUE would be very different.[5] Here are two examples:

Organization One: SoC = 50; Layers = 3; Total managers = 21; Total ICs = 1000; PUE = 1.35
Organization Two: SoC = 5; Layers = 5; Total managers = 251; Total ICs = 1000; PUE = 1.51

Organization One is very flat. To accommodate 1000 ICs with a SoC of 50, we'd need 20 leaders, each of them would manage 50 ICs. The CEO would then have 20 reports, which is less than the allowed SoC. Managing such a large group of ICs is very hard. It is likely that management overhead for these 20 managers would be so high that their VPW percentage would be even smaller

---

5. Please see a visualization of the PUE for different organizational structures and VPW configurations at https://digitalagencybook.org/visualizations/org-tree.

than 30% on average, as they would have no time to do any technical work. A thankless job indeed. Additionally, if we adopted this flat structure, we wouldn't benefit from too much gain in PUE from our original, and more conventional, four layer design with a SoC of 10.

Organization Two limits the SoC to five. To accommodate one thousand ICs, we'd need two hundred first line managers. In turn, the other layers, viewed from the bottom to the top, would be 40, 8, 2, and 1, leaving us with 251 managers. In this scenario the PUE looks very bad, as we have so many managers in relation to the number of ICs. However, the two hundred line managers probably have more free time on their hands, since their team is limited to five ICs only, and may have more time for some VPW. If we increase their VPW percentage to 50% from 30%, the update PUE becomes 1.42. Much better, but still not worthwhile compared to the more standard design of four layers and SoC of 10. Another problem with this organization is its increased number of layers.

Ben Horowitz, the author of *The Hard Thing About Hard Things*,[6] emphasizes how communication is crucial for organizations, stating "perhaps the CEO's most important operational responsibility is designing and implementing the communication architecture for her company." The more layers an organization has, the more complex the communication architecture. This is another

---

6. *The Hard Thing About Hard Things: Building a Business When There Are No Easy Answers*, Ben Horowitz, Harper Business, 2014.

downside of limiting the SoC and increasing the number of layers. Broken communication paths would likely reduce the amount of VPW, especially for the ICs at the bottom of the hierarchy, driving the PUE even higher.

The split between managers and ICs is crucial for a few reasons. First and foremost, technical VPW needs to be done by technicians, not by managers. Additionally, the skills required by the two roles are very different. The Peter Principle assumes that, by going up the hierarchy, what you learned in your previous role will not help you as your new job would be so different. This is not the case if we cleanly separate the management and IC tracks. Most tech companies have already realized this. Two relevant books on this subject are *The Manager's Path*[7] and *The Staff Engineer's Path*,[8] which describe the challenges and provide career advice for tech managers and ICs, respectively. Unfortunately, there are still organizations that convert talented engineers to managers, sometimes prematurely. And in many cases, this conversion is wrongly perceived as a promotion instead of a career change.

Having well-defined career paths for management and technical tracks is crucial. Without that, strong ICs may view no path forward for career progression. At Microsoft, Google, Amazon, and most of the big tech

---

7. *The Manager's Path: A Guide for Tech Leaders Navigating Growth and Change,* Camille Fournier, O'Reilly Media, 2017.
8. *The Staff Engineer's Path: A Guide for Individual Contributors Navigating Growth and Change,* Tanya Reilly, O'Reilly Media, 2022.

companies, this separation is clear and there are prominent role models on both tracks. If there were no prominent ICs and every newcomer that joined Microsoft wanted to be a manager, we'd have a problem. There are very few manager roles, especially at the top of the pyramid. With SoC equal to ten, there are ten times more ICs roles than manager roles. In most big techs, strong ICs can rise up to be very influential and be at the same seniority level as big shot vice presidents.

By having a strong IC culture, in which newcomers have the conditions to progress in their careers, regardless of the path they choose, with the support from mentors and a great working environment, we increase their productivity and therefore their average percentage of VPW. Another important part of increasing VPW is leveraging platform investments. As we've seen in the last chapter, cloud computing lifts a lot of the burden from engineers, which can now focus on the differentiating services their companies provide, while relying on the cloud's resilient and scalable computing infrastructure. Similarly, the word processor increases the productivity of writers and web search increases the productivity of academics. Platform leverage is a crucial mechanism for achieving higher percentages of VPW.

We are in the age of AI and there is debate of how it will impact productivity and the job market. We'll go over AI's impact in more detail. But first, I'd like to go deeper into organization design with a few examples of companies I've worked for.

Too many or too few managers

When I worked at Google, it was a very flat organization. I joined the company in 2011. The SoC was still high and we had very few layers. In the early days, it was even more extreme, with a few engineering managers having teams of more than 50 ICs, as in the case of the fictitious Organization One in our example. Maybe this is an urban legend, but I heard that an early employee quit and didn't properly communicate his departure. It took three months for people to realize the guy was no longer there.

The engineering managers back then were all very seasoned. However, managing over 50 ICs, they had very little time to do any technical work. They could not even track down that one of their team members had quit! Although they played an important role in hiring, these managers did zero VPW. But by ensuring they hired only top talent, Google built a very strong engineering culture. They hired only the best ICs, and gave them a lot of freedom. I once asked Nadav Eiron, who was my manager while I was at Google, how they made sure people are productive, given there was so little management oversight. His response was something like "we only hire motivated people. We trust that they will either find a way to be productive or realize that Google is not for them and quit."

That focus on hiring the best and "letting the flowers bloom" was crucial to Google's success. Besides hiring, another key factor was their platform investments. They built several platforms that greatly improved engineering productivity and that provided a common framework for collaboration. One example is the platforms that allowed developers to view and modify any code for any of Google's services, while enforcing strict guidelines and promoting engineering rigor. By providing freedom and great platforms to talented ICs, magic happened.

This is easier said than done. Nassim Taleb claims Google's success was a black swan event. It was a rare combination of the right people, leveraging the right platforms to solve the right problem. If any of these factors were absent, the story would have been different. Other companies had talented people, but didn't have the right platforms. On the other hand, other companies may have had the right people and platforms, but tried to solve the wrong problem. As we've discussed, online advertising is a goldmine that has allowed Google to reinvest even more in people and platforms.

Keeping a SoC of fifty, hiring the best ICs, and letting the flowers bloom isn't a formula that can be easily replicated. Even Google moved away from that model to a more conventional organizational structure with more layers and smaller SoC. It is also hard, for any company, to rely on only hiring top talent. Successful companies must have a strategy to grow talent from within, and that is only possible with a good management team and well defined career ladders for managers and ICs.

That was not the case when I joined Stone,[9] a fintech company that operates in the Brazilian market. The

---

9. https://www.stone.co/en/ (Last accessed in March 2024).

company grew very fast after its successful debut on Nasdaq in 2018. When I started as chief technology officer (CTO) in 2022, it had over ten thousand employees, two thousand of those in engineering. Still, despite being one of the largest technology companies in Brazil, it didn't have formal career ladders. And we had an aggravated case of Peter's Principle. Talented young developers were quickly converted to become managers. I carefully chose the word converted and not promoted here, as I strongly believe that switching from the IC track to the manager's track, or vice versa, is not a promotion. It is a career transition.

In Stone's case, it was seen as a promotion, due to the lack of a formal career path for ICs. The company ended up with inexperienced managers. And lots of them. In 2022, five hundred out of the two thousand employees were managers. The average SoC was four, with many managers having a single employee reporting to them. The company was an even more extreme case than Organization Two in our example above, which had a PUE of 1.51. As the managers were inexperienced, even though they were managing small teams, their VPW percentage was still low. If we optimistically assume a VPW of 30% for managers and 70% for ICs, their PUE was around 1.67.

A lot of my work as CTO was to make the organization more efficient. I tackled multiple areas in parallel, including the following:

1. Definition of the IC and management career ladders. One of the first things I did as CTO was to define

the career ladders for ICs and managers. These ladders were largely based on what most tech companies in Silicon Valley have adopted. There is parity of levels between ICs and managers in every level, except for the first IC levels that have no manager equivalents, as we require managers to be experienced. Everyone has to start as an IC. It is hard to coach basketball, or any sport, if you never played yourself. The same is true for management. Managers that have never been ICs are unable to understand the details of the work. They would have difficulties, for instance, estimating the level of complexity of individual tasks. This type of career ladder is normally referred to as Y, as it starts with a common base and bifurcates into management and IC tracks. Figure 10 shows the one we adopted at Stone. We also made Stone's ladder public, to help other companies going through the same transition.

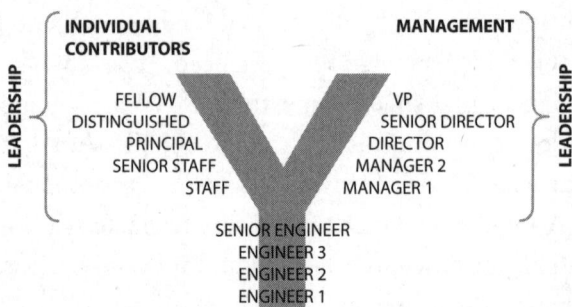

Figure 10. Y-career ladder. The initial branches are IC-only. The leadership tracks start at the base of the Y, with parallel tracks for ICs and managers.

2. Increased the level of seniority. Inexperienced managers are not able to properly mentor and evaluate their direct reports. Moreover, if they don't know how to handle a situation they may need help from their own manager, further increasing management overhead and reducing the organization's VPW. Experienced managers increase organizational efficiency. At Stone, we decided to increase the seniority level of the entire technology team, especially the managers. Having more experienced managers is a prerequisite for us to be able to increase the team sizes. Having the career ladder in place, we shifted our hiring strategy to focus on more senior positions. Gradually, we were able to increase the level of seniority of the tech team. This had a compound effect, as talented and experienced employees attract other talented and experienced employees.

3. Increased the SoC. With better managers, we could increase the SoC and have an organization with larger teams. We were able to gradually change SoC from four to a better number between six and eight. We started with some low hanging fruit. Every manager that had an SoC less than three had their teams collapsed. And every manager that was very junior, in the base part of the Y-career ladder, became an IC. We didn't blindly apply these rules. We had to analyze case by case as there were exceptions. Sometimes a team had only three people because it was still expanding, or because it handled a very technical and specialized part of the business. Converting managers to ICs is also challenging. Sometimes they were

not able to, or were not interested in, doing engineering work and we had to let them go, or they chose to leave. When I left the company in 2025, we were still not done with our efforts to increase the seniority level and SoC.

4. Invested in manager training. Besides hiring more experienced managers, it is important to continuously invest in manager education. This is always true, but even more so in Stone's case given all the changes we were driving. We were shifting the company from a start-up, focused on time-to-market, to a more structured company, focused on the long term. "It's not a sprint, it is a marathon" became my mantra, and in addition to being the CTO, I also played the role of the CRO, chief repetition officer. Cultural change takes time and effort. And it is paramount to have all the managers on the same page, singing the same song. One of the worst things that can happen is for an employee to ask the same question for two different managers in the organization and receive conflicting information. Training not only increases the manager's ability to remove friction and empower the ICs, it also brings consistency to their message. An unified leadership team should work as a deterministic algorithm, always producing the same results when given the same inputs.

5. Reduced management overhead. More senior and well trained managers will naturally deliver more VPW. But in addition to focusing on hiring and providing training, we also took a closer look at Stone's management processes. The annual performance evaluation cycle, for instance, was very time consuming. It had many phases,

including detailed forms with lots of non-optional fields. Moreover, as the managers were inexperienced, this time consuming process didn't even generate the desired outcomes. A typical case of **garbage in, garbage out** (GIGO). The managers wrote poor evaluations and the process didn't properly differentiate the employees. We greatly simplified the process, making it less onerous. Besides being simpler, the new process worked better, providing more differentiation between the top performers from the middle of the pack and the low performers. In addition to performance management, we looked into all the processes that added overhead, such as planning and budgeting, and strived to simplify them.

6. Invested in platforms. Perhaps the most important change I helped initiate at Stone was investing in platforms. Once I arrived, teams were organized by product. Our solution for small and medium merchants had been developed completely independently from the one for micro merchants. These were two completely different apps, which used distinct backends and were developed by independent teams. While this structure gave independence to the teams, it led to a lot of duplication and a bloated organization. It worked well when the company was smaller and had to focus on time to market. Teams were empowered to work independently on new features to capture market share. However, as the company grew, we needed the different products to work well together, providing a frictionless experience to users and enabling collaborations between teams.

By shifting to organize by platforms, instead of by product, we pivoted the existing teams, "product SMB" and "product micro," into infrastructure, data platform, payments platform, banking platform, and so on. Figure 11 illustrates this change. The advantage is that these platforms now were leveraged by multiple products. The same banking platform was used by the SMB and micro products. We called this strategy "build once, use many." This was a long term change. Doing this transformation requires resolving duplication and unifying backends, which is not trivial to do, especially for Stone, which has millions of active customers. However, it leads to a more efficient organization and a more consistent product experience.

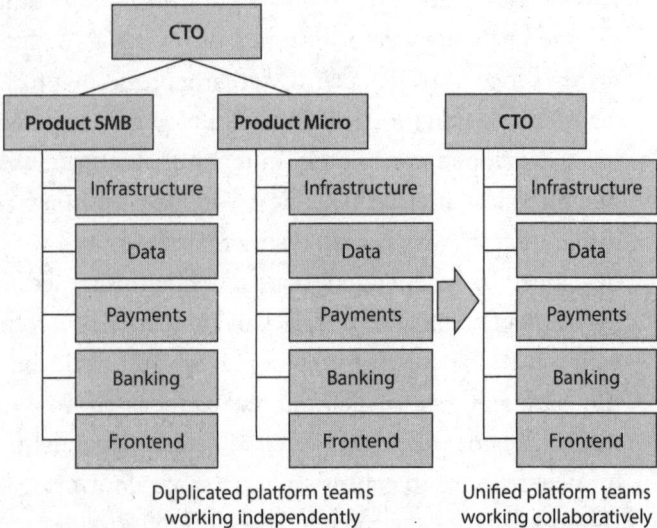

Figure 11. The organizational change at Stone transformed a product-oriented structure into a platform-oriented structure.

The other important aspect of this change is that it required that teams start collaborating. When "product micro" was an independent team, with complete autonomy, it could modify its backend systems without any coordination with other teams. In an organization by platforms, this is no longer true. Any changes in the banking platform would require collaborating with that team. That is only possible when managers promote a strong engineering culture, with well-defined guidelines and processes. As I wrote in *A Platform Mindset*, platforms are the backbone of a technology company, but so is collaboration.

Both examples, Stone and Google, are extreme. Too many managers is unneeded overhead. It is like too much cooling in a data center. Not needed and wasteful. You want just enough managers and cooling infrastructure to keep people and servers working happily. The opposite is also problematic. Too few managers doesn't scale. Google's success was a black swan event. Even they changed their structure over time, to manage their growth. Next, we'll discuss in more detail how platforms can be used to address the issue of scaling the organization.

The multiplier effect of leaders and platforms

In her book, *Multipliers*,[10] Liz Wiseman describes how leaders can amplify or reduce the impact of their organization. It is possible to make an organization of one

---

10. *Multipliers: How the Best Leaders Make Everyone Smarter*, Liz Wiseman, Harper Business, 2017.

hundred employees to behave as a team of two hundred or as a team of fifty. Leaders who amplify the impact of their teams are **multipliers**, while the leaders that limit the ability of their organizations to deliver value are **diminishers**. Although Liz Wiseman's book focuses on managers, the concept applies equally to senior ICs.

Experienced IC developers who spend most of their time complaining about everything, the quality of the code, the stupid CEO, and the bad food in the cafeteria, are diminishers. If they are influential in the organization, they will negatively impact morale. We all know these "brilliant jerks," and they are toxic to the organization. On the other end of the spectrum, the IC developers who are always helpful to others, mentoring junior engineers and providing helpful comments during code reviews, would be multipliers. They would be using their knowledge and experience to empower others to achieve more.

Once upon a time, I had a direct report who was an archetypical diminisher. When I started managing him, he was leading a very large organization. As I tried to understand why his team was so bloated, I soon realized that he had built entire platforms under him, even when these were available elsewhere in the company. To avoid taking dependencies on other teams, he preferred building new systems from scratch, so he could "have more control." Over time, his services became so complex that every new change, even simple ones, required an oversized investment, both in the number of engineers and in time. The aggregate VPW for his organization was very low and every project took longer than it should.

Reversing a diminisher's impact is a significant challenge. In this case, in addition to organizational changes, it required platform deduplication that took years. This is work similar to the one I had to do at Stone, when I pivoted the teams to be organized by platforms. We had to invest in deep technical work on platform convergence. Despite being hard, I'm a big believer that investment in common platforms pays off, as they too are multipliers.

A very drastic example of how platforms can greatly increase productivity was the use of digital computers, such as the ENIAC, to replace human computers. The ENIAC performed calculations ten thousand times faster than humans.[11] A single person could program the ENIAC and produce the same amount of VPW as an organization of ten thousand human computers. Most platforms are not as impactful, but still, leveraging platforms is one of the most effective ways to increase VPW.

Electronic health record (EHR) platforms allow doctors and nurses direct access to patients' medical history. EHR not only increases the productivity of health care providers, it also decreases the chances of medication and treatment errors and streamlines the health care facility workflows. Some studies show gains in productivity of 5-10% for doctors, allowing more precise patient visits,

---

11. https://www.historyofinformation.com/detail.php?id=636 (Last accessed in March 2025).

reduction in documentation time by nurses in the order of 20-30%, and shorter wait times for patients.[12]

There are examples in every domain, such as the use of robotics in Amazon's warehouses, precision agriculture systems for farming, and cloud computing for online services. I'm a believer in technology that enhances human experiences, including how we work, and not in technologies that enslave us. Surveillance and timekeeping systems for retail workers, for example, add little efficiency gains and degrade the employee experience, as described in *Power and Progress*. The examples I gave are systems that automate repetitive work and empower employees to focus on the creative parts of their jobs. Each of these platforms has their own multiplier effect, and their adoption must be accompanied by organizational changes that take the VPW improvement into account. The same is true when replacing a diminisher leader by a multiplier. The organization will become more productive, but it may require changes. For instance, the new leader may decide to take dependency on a platform developed by a sister team and reassign the employees working on the duplicated system to explore a new product idea.

When we think about optimizing an organization that has become more productive, due to a change in leadership or the adoption of a new platform, the first option that

---

12. https://www.ambula.io/how-does-ehr-increase-productivity/ (Last accessed in March 2025).

may come to mind is reducing personnel. Although this is the more straightforward option, and I want to spend some time going over it, there is another possibility that we should consider. Reinvesting the gains in innovation.

Who has time to think about the future?

One of the most influential and insightful business books of all time is Clayton Christensen's, *The Innovator's Dilemma*.[13] The main idea, which may seem surprising, is that well-managed companies that pay attention to their customers and build solutions to address their immediate needs, will eventually fail. And the more they dedicate resources to listen to their customers, and be "customer-obsessed," without paying attention to technological advances that may be imminent or already happening, the more they risk failing.

Amazon has become a symbol of customer centricity. Several of its processes, described in *Working Backwards*,[14] start from and strive to deliver value to their customers. The PR/FAQ, which stands for press release and frequently asked questions, is one of these processes. It mandates that before developing any new product or feature, the responsible team should develop a press release (PR) highlighting the impact it will have

13. *The Innovator's Dilemma: When New Technologies Cause Great Firms to Fail,* Clayton Christensen, Harvard Business Review Press, 1997.
14. *Working Backwards: Insights, Stories, and Secrets from Inside Amazon,* Colin Bryar and Bill Carr, St. Martin's Press, 2017.

in the lives of potential customers, and an accompanying document anticipating their frequently asked questions (FAQ). Until a PR/FAQ is written, discussed, and approved, nothing new gets built in Amazon.

There are countless stories of how Amazon's customer service employees go above and beyond to make their clients happy. For example, when a customer got a damaged product before Christmas, the Amazon rep not only expedited the replacement for the item but also sent the customer a gift card.[15] In my many years being an Amazon customer, I only have had one issue. Customer service didn't want to refund me for the difference between the treadmill I bought for my mother and the one they mistakenly shipped to her. I used the magic phrase, "well, thank you for being customer-obsessed and trying to resolve my issue." I was being sarcastic, but it immediately triggered an escalation. And I got to talk to a manager who quickly solved the issue.

What may not be so obvious, is that Amazon is not only customer-centric, it is also efficiency-oriented and innovative. By optimizing its operational costs, the company is able to run its current businesses at the lowest cost possible. This includes its optimized supply-chain, warehouse automation, and its focus on organizational efficiency. By adapting technologies like robotics and cloud platforms,

---

15. https://www.sobot.io/blog/behind-amazons-success-customer-service-strategies-and-innovations/ (Last accessed in March 2025).

Amazon is able to deliver more VPW, in other words, it can do more with less. And the value it produces by being "frugal," is invested in innovation. It could, instead, run leaner and leaner, sizing its teams to the minimum number of employees required to maintain its operations running.

Without investing in innovation, Amazon would risk its leading position due to two forces. The first is **deterioration.** With a team sized only to maintain its current operations, there would be no available resources to investigate new technologies. This means that their currently optimized VPW percentage would, eventually, no longer be best in class due to technological innovations that it would not be taking advantage of. Let's say, if no one in Amazon had the responsibility of investigating how AI can impact software development, its tech teams would be stuck with current tools. Over time, they wouldn't be as productive when compared to other companies. They would no longer be operating at an optimal VPW. Moreover, assuming these new technologies are enriching to the employee experience, the team should be eager to adopt them. If they don't, the company may no longer be attractive to new engineers, who would see them as the dinosaurs of software development. A team with subpar talent also negatively impacts VPW.

The second force is **disruption**. With a team sized to focus only on the current business, there is no way the company can be innovative. This means no new business areas will be created and new products and services will not be developed. Even if the company has a very robust business, after time, new technologies may make that business less relevant or obsolete. This is the main thesis of the late Clayton Christensen's "the innovator's dilemma." Amazon being completely focussed on being the largest and most efficient book retailer in the world, could have missed the e-book and web services, had it not prioritized and funded the development of new technologies. A company that only cuts costs and doesn't invest in innovation is racing to the bottom.

One of the examples Christensen used to describe the innovator's dilemma is the automotive industry. American car manufacturers, including General Motors, Ford, and Chrysler, were the market leaders in the fifties and sixties. They catered to their existing customers, who wanted ever bigger and luxurious cars. The Japanese automakers entered the U.S. market around that time, with compact, fuel-efficient cars. Being inexpensive, the Japanese cars were appealing to first-time buyers, but were considered inferior to the American luxury models.

The 1973 oil crisis was caused by a total oil embargo by the Arab oil-producing nations against countries that had supported Israel at any point during the 1973 Yom Kippur War.[16] It caused oil prices to skyrocket, accelerating the demand for fuel-efficient cars and American automakers were too slow to react. They were **disrupted** by the technology of fuel efficient cars. It is not that they

---

16. https://en.wikipedia.org/wiki/1973_oil_crisis (Last accessed in March 2025).

didn't know about these cars, they simply decided not to focus on them, as they were happy with their more profitable, luxury car market. And when they did, as in the case of the Ford Pinto we covered in the previous chapter, they tried to catch up by accelerating production times and made too many mistakes.

By the eighties and nineties, Toyota, Honda, and Nissan were no longer seen as the makers of cheap, low quality cars. They were leaders in efficiency and reliability. By the 2000s, they had captured much of the American market. There are many other examples that highlight the importance of continued investments in innovation. When incumbents get complacent and do not reinvest their profits into new areas, they risk being overtaken by disruptors. Microsoft CEO, Satya Nadella, says he would like his technologists to "see around the corner." This means understanding and investing in new and promising technologies.

By focusing on organizational efficiency, we can have fewer resources applied to operating our current business. If we don't want **deterioration** and **disruption** to occur, we should reinvest our efficiency gains in innovation. To combat deterioration we must invest in adopting or developing technologies that will improve our productivity, creating a virtuous cycle. To combat disruption, we must invest in new technologies that will open new market opportunities, be it the kindle, robotics, or fuel-efficient cars. But even with proper investments, innovations don't simply happen. They take time and concentrated effort. Some directions may fail, and we may need to pivot. Sometimes we may feel we are stuck. What are the strategies to effectively enable innovation?

Why nothing works

Figure 12 illustrates the virtuous cycle we just talked about. Investments in productivity will optimize the bottom line. We'll be able to support our current products with less. If part of the gains are reinvested, we can explore new market opportunities. When those are successful, they will improve the top line. By reinvesting part of the new gains, we can further improve productivity. And we can keep repeating this cycle until the company is very successful and we are very old. A simple formula for success. But this is much easier said than done.

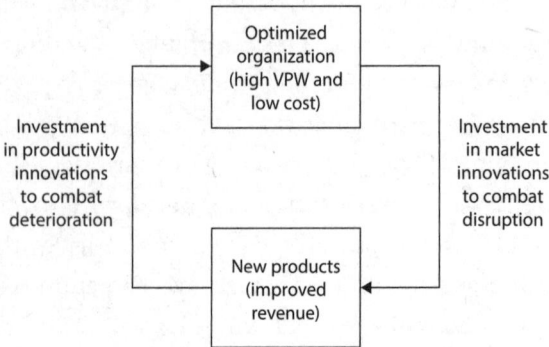

Figure 12. The virtuous cycle of innovation. Innovation in productivity improves the bottom line while market innovations improve the top line. Continued reinvestments of these gains combat deterioration and disruption. By investing in innovation, the company will attract better leaders, both managers and ICs, who will further increase productivity and produce novel ideas.

Hiring a chief strategy officer (CSO) to dictate which innovative projects the team should pursue normally doesn't work. Most strategists are disconnected from the work being done on the ground by the teams. A lot of innovation happens the other way around, bottom up. Technical ICs working on projects are the ones with a better chance inventing new methods to boost productivity and new features that could lead the company to embrace completely new market opportunities. The issue is that we need mechanisms for these ideas to flow upwards, from the ICs to the company leadership. These ideas need to be evaluated, and the good ones should become funded projects.

The balance between top-down directives and the bottom-up empowerment of ICs is not unique to business. In his new book, *Why Nothing Works*,[17] Marc J. Dunkelman provides a very interesting interpretation about the American government's inability to execute large scale projects. His main thesis is that progressives are torn with two conflicting ideals, which he defines as Hamiltonian and Jeffersonian. The Hamiltonian impulse advocates for centralized power, emphasizing the role of a strong government, led by experts to drive societal change. The Jeffersonian impulse, on the other hand, advocates for the protection of individual rights and the decentralization of power to local entities.

---

17. *Why Nothing Works: Who Killed Progress–and How to Bring It Back,* Marc J. Dunkelman, PublicAffairs, 2025.

According to Dunkelman, both impulses have always been present in the progressive movement, with each being more prominent at different times. Until the end of 2024, when he wrote the book, we were leaning towards Jeffersonianism, with so many checks on authority that we are unable to get anything done. One of the examples cited in the book is the prolonged renovation efforts of Penn Station in New York City. Once an architectural landmark, Penn Station was demolished in the sixties. Despite widespread agreement on the needed reform, the project lasted more than three decades, with many delays and setbacks. Dunkelman compares this project with the many large infrastructure projects executed by urban planner Robert Moses in the mid-20th century. Although there were controversies, Moses was able to execute these projects swiftly.

The two impulses are always present in the minds of progressives. We want to execute on climate change, but also don't want police abusing citizens and construction projects displacing entire neighborhoods. Progressives must acknowledge and reconcile these two valid but contradicting ideals if we want to get anything done. As Dunkelman puts it, "there is no way to serve the greater good without exacting some cost on at least someone." We must establish cooperation and trust to minimize such inevitable costs to drive progress.

Back to the corporate world, when the flow of ideas is not supported, innovation doesn't happen and the virtuous cycle shown in Figure 12 stops. Let's consider two extreme scenarios.

- A Hamiltonian, top down-organization: when the CSO, chief strategy officer, sitting in the glass office of the company headquarters, defines the projects an organization will take on, ICs that work on the ground will feel demoralized. They are the ones that have a better understanding of the intricacies of the problems they are working on.
- A Jeffersonian, bottom-up organization: when the ICs are properly empowered, but there is no central oversight, there is a chance that the organization will produce many non-cohesive solutions, instead of a product that is aligned with the company mission. This may work for a while, when the company is small, but it does not scale.

Most companies are primarily Hamiltonians. The C-Suite defines the direction, which is carried out level-by-level down the hierarchy until reaching the ICs, who will promptly execute orders. When the market fluctuates and directions change, a new set of orders will go down. ICs are then expected to ditch everything they are doing, and promptly start on the new projects. After all, this time around the executives are right and these new projects will cure world hunger. Over time, ICs get frustrated and leave. This is not a problem, as the Hamiltonias view them as expendable. Just hire new ones. Many companies even optimize to extract the most of ICs in the first few years, demanding a lot of work for little pay, knowing they will eventually leave for a better place.

Google in the early days was a Jeffersonian organization. It hired the best ICs in the industry and let them work, with little oversight. They developed innovations like PageRank and anchor text, and disrupted all the existing search engines at the time. And their investment in infrastructure, required to run search at a global scale, allowed them to spawn new, successful businesses. Over time, however, many of the Google engineers were not working on search. They were working on a multitude of projects, not necessarily aligned with the company mission. This was exacerbated by the 20% rule, which allowed everyone in the company to spend 20% of their time working on personal projects. Eventually, Google decided to "defragment" these projects. They also eliminated the 20% rule, and installed more engineering layers and more company oversight.

The ideal organization achieves a healthy balance between **empowerment** and **oversight**. The heart of the organization are the strong technical ICs. They deeply understand the problem space and the technology, and can drive meaningful change. They must be empowered. However, without any central guidance, they may produce solutions that are not synergic with the company direction. I once worked with engineers that loved a given programming language. They were more interested in convincing leadership that we should rewrite all our systems using that language than in addressing the backlog of customer issues or working on platform improvements.

Good ICs are strong minded. They have opinions about everything and they are always right, or at least they are very good at convincing you that they are right. All the time. That is why oversight is so important. There are multiple programming languages, all of which have their pros and cons. It is important to align in one, or a few, depending on the complexities of your organization, and drive standardization across the organization, than to let each IC have their own way. Common platforms and tooling brings synergies. In this example, a single programming language enables developers across the organization to collaborate in projects together without having to learn new tools. But how to drive these decisions?

40-60 and 10-90 decisions

It is not a good idea to let opinionated ICs pick their own programming language or let the CTO issue a top down mandate without listening to the team. One of the lessons I learned from the late Google engineer Luiz Andre Barroso is the 40-60 rule. Luiz used to say that there are two types of decisions leaders are asked to make on a regular basis, 40-60 and 10-90. 40-60 decisions are those where one option is slightly better than the other. Choosing C# as the company-wide programming language would require an investment of 40%, while if we adopted Java it would have been 60%.

This is just a fictitious example. I'm a C++ dinosaur, and I don't have a strong opinion about this. And that may be exactly the point. Both languages will have their strengths. Converging in one of them is much better than not converging in any, independently of which one is chosen. 10-90 decisions, on the other hand, are much more impactful. The worse option is much worse. Nine times worse. Amazon classifies decisions as single-way and two-way doors. One-way door decisions are irreversible, or very hard to reverse, and have significant consequences. Two-way door decisions are simple to reverse if they don't work out, requiring less scrutiny.

In *Working Backwards* they use the Kindle as an example of a one-way door decision. Deciding to implement the Kindle required starting a new hardware division at Amazon. Something costly and that cannot easily be reversed. They had to invest in R&D and build a completely new ecosystem of e-books, publishers, and content delivery. That is a much more serious decision than changes to the Amazon website. These are considered two-way door decisions, as they can be easily reversed. I still prefer the 40-60 and 10-90 definition over one-way and two-way doors, as it more explicitly describes the concept, and represents the non-binary nature of decision making.

Although the 40-60 decisions are less consequential to the company, they may still be a matter of life and death for some of the ICs. It is very important for leadership to realize that some of these decisions may appear as limiting or restrictive. Imposing the use of C# throughout the company may largely impact developers who spent most of their day writing, testing, and reviewing code. Especially if they previously had the freedom to choose whatever language they wished. So, although the chosen language itself may not matter, how we make the decision matters.

The solution is to involve all the senior leaders, both ICs and managers, who may have relevant opinions in the decision making process. It is impossible to include everyone in the company, but senior leaders should be aware of the different opinions and should represent the conflicting positions. How much time is spent in each decision depends on its importance and how controversial it is. A 10-90 decision may be more straightforward than a 40-60 decision. It is important to dedicate as much time as needed to make sure all opinions are heard. If the process is data-driven and includes all the relevant counterparts, we have a much better chance that the organization will accept the decision.

After a decision is made, it will be seen as part of the culture and the environment. Old timers who had to endure the transition to C# may have perceived it as a loss of freedom. Everyone that joined the company after the decision was made and implemented will not see it as a restriction or a loss of freedom. In their mind, this is how the company operates. They may or may not believe in the importance of having a common language across the company. And they may or may not like C#. But independently of that, that is simply how things work in the company they just joined. If they are gravely allergic to C#, they hopefully asked which programming languages were adopted before joining the company.

An inclusive decision making process helps us balance the Hamiltonian and Jeffersonian impulses within the organization. The CTO may centrally decide that the company should adopt a single programming language, providing her compelling arguments. Instead of imposing this mandate top-down, she would work with her leaders and the strong technical ICs, empowering them to discuss the pros and cons and propose the solution. I did exactly that when I was Stone's CTO. The team came back with a proposal recommending three languages instead of one. One for backends, one for data manipulation, and one for embedded systems. The overall outcome was much better than if I had pushed it top down, in part because technical ICs were involved in the decision making process, and played a fundamental role in driving its execution.

Of course, having the right decision making process does not guarantee that we'll make the right decisions. If the decisions themselves don't make sense or if the decision-makers don't properly analyze the data, there is no magical process that will save us. This brings us to the next topic. We need good people. Not only to drive decisions, but for everything we do.

Attracting and keeping the right people

A data center may have a good power usage efficiency, PUE, and the servers can be well-utilized, but if the applications running in these servers are not doing anything useful, no value is being produced. In the same way, an organization may have a good personal usage efficiency, our new definition of PUE, and good guidelines, but if no one is working on relevant projects, no true value will be produced. We need people with good technical judgement to set the direction of the organization, and strong ICs to drive execution.

It is widely known that good people attract good people. The opposite is also true. This is why the Silicon Valley talent search is so fierce. Top talent will not only have the multiplier effect we already talked about, they will also be role models for the rest of the organization. I firmly believe ICs are the heart and soul of the organization. Therefore, one of my main goals as an engineering leader is to drive a strong IC culture and to have prominent ICs leading important and visible projects.

It is true that Satya Nadella is a very charismatic and a very inspirational role model. However, if every developer joining Microsoft views his role as their dream job, they are putting themselves in a very difficult position. The top of the organization is a pyramid and there are very few roles for senior leaders. And there is only one CEO. If, on the other hand, the newcomers joining Microsoft are excited about building technology, there are many more spots as senior ICs in the company. Therefore, it is very important that organizations have technical IC leaders as prominent role models that are seen as being very influential in the organization.

Hiring top talent is hard, so it is disappointing when good people don't produce to their fullest potential. Or worse, when they become demotivated and leave the company. This will always happen to an extent. People are different and their interests and circumstances change over time. The managers' job is to counter that. They should spend a substantial amount of time building an environment where employees can prosper. This is the most valuable, non-VPW that managers do. Sometimes, despite our best efforts to keep the hiring bar high, we'll make mistakes and hire employees that won't work out.

The candidate may be smart and motivated, but hard to work with, or not aligned with the company mission and culture. Or she may have forgotten to say that she only likes to program in Lisp and be furious when she realizes that her new company has other guidelines for software development. Whatever the reason, I always like to see the relationship with a new employee as a journey that resembles more a marathon than a sprint. In a marathon, there can be stumbles along the way, but with the right support from your teammates, it is possible to be successful in the long run. I'm much more inclined to measure progress over time than to have a short-term view of someone's contributions. If the employee is not learning and improving overtime, we have a much stronger signal that we made a hiring mistake than if he or she didn't perform well in their first project.

More than once, I had employees that joined my organization from different backgrounds and took a while to become productive. But they were very motivated, open to feedback, and showed steady progress. Employees, especially newcomers, often ask me what they need to do to be successful at work. My answer is always the same: have a growth mindset. The concept of growth mindset was proposed by Carol Dweck in her seminal book *Mindset: The New Psychology of Success*[18] and it was one of the key pillars of Microsoft's cultural transformation led by CEO Satya Nadella. Employees that are open to take feedback, face their own limitations, and learn how to overcome these limitations to do their work better, will perform very well in their jobs. Even if they have a lot to learn in the beginning, they will close these gaps in the long run. That's why viewing performance management as a marathon is so important.

Employees that have a closed mindset, are those that prefer to go to the dentist than to listen to feedback. Even if they are really smart and hardworking, they may be doing work that is not aligned to the company mission. If they don't internalize that they need to change, they may behave like a computer that is running a program to count the stars in the sky. They are really busy, but nothing good will come out of it. Identifying employees with a close mindset as soon as possible is important, as it is unlikely that they will improve their performance overtime.

---

18. *Mindset: The New Psychology of Success,* Carol Dweck, Ballantine Books, 2007.

Throughout this chapter I've made subtle parallels between humans and computers, such as aligning the definition of personal and power usage efficiency, and how bad performers relate to pointless applications. Although I don't think that servers and humans are exchangeable, drawing this parallel was intentional, given all the buzz about AI and how it will disrupt the way we work. That threat seems especially real for knowledge workers. We'll cover AI in detail in the next chapter, but I'd like to start by discussing some of its implications to productivity.

The digital-only myth "AI"

Gabriel Garcia Marquez, affectionately known as Gabo, spent eighteen months of full-time writing to produce the tale of the city of Macondo and the Buendia family described in *One Hundred Years of Solitude*. During this time, his wife Mercedes had to use credit for groceries and even rent. After exhausting their savings, they tried to pawn Mercedes' jewelry to finance the writing, only to discover that the supposedly precious stones were all glass. They eventually pawned their family car. Still, when he was done, Gabo had no money to mail the whole manuscript to his editor. He and Mercedes split it into two parts, as they had only enough to ship half of the book. By mistake, they mailed the second part instead of the first. Luckily, before they could collect enough money to mail the beginning of the story, they got the shipping money from the editor, who was by then eager to read it.

Gabo wrote the book between 1965 and 1966. The story for writing and publishing the book may be worth a new book in itself, especially due to the financial difficulties that the family had to endure during the time, and how Mercedes was able to overcome it and make sure there was food on the table every day. Gabo said in a speech "it seems incredible, but one of my most pressing problems was finding paper for the typewriter," and that he had "the bad manners of believing that misspelled words, language mistakes, or errors in grammar were actually created." When detected, he would "tear up the page and throw it in the trash basket to start again."[19] We will never know how much of this self-imposed "quality assurance" program improved the book, compared to what would have happened if Gabo had modern technology at his disposal. We do know for sure that it would have been easier, less costly, and faster to email the manuscript to the editor rather than sending it by snail mail in two parcels, had email been invented a few decades earlier.

Les Earnest developed the first computer-assisted spell checker in 1961, at Stanford's AI Lab (SAIL).[20] His program didn't suggest corrections. It simply checked each word against a fixed dictionary of ten thousand acceptable words. In 1971, a few years after Gabo finished writing *One Hundred Years of Solitude,* Earnest's student

19. https://www.youtube.com/watch?v=pzx1msp4-Ik (Last accessed in March 2025).
20. https://en.wikipedia.org/wiki/Spell_checker (Last accessed in March 2025).

Ralph Gorin wrote SPELL, the first true spelling correction software. The program searched its dictionary for corrections of single letter misspellings or adjacent letter transpositions, such as "hte" instead of "the." Since this program was only looking at a word at a time, transpositions present in the dictionary, such as "form" instead of "from," slipped through undetected.

Word processing technology has evolved a lot since then, especially after the popularization of personal computers in the eighties. Starting with integrated spell checkers, featuring larger dictionaries and built-in grammar correction, in the mid eighties and early nineties. Soon after that, Microsoft Word introduced AutoComplete and AutoCorrect. AutoComplete provided context-aware word suggestions, while AutoCorrect automatically fixed the spelling of words as you typed. By using context, correcting "form" to "from" and vice versa now became feasible.

During the early 2000s, not only web search became popular but we also saw great advances in automatic translation. In 2006, Google Translate was released, providing real-time translation across dozens of languages. While the early translation software was based on rules and dictionary lookups, Google Translate was, instead, based on statistical machine translation (SMT). SMT-based systems use vast amounts of bilingual text to "learn" new translations. While in the previous systems the rules were fixed, SMT improved over time, getting better when provided with new books and their translations.

By the 2020's came the first commercial systems with context-aware, whole sentence suggestions, such as Google Smart Compose. And since the popularization of OpenAI's ChatGPT and Microsoft Copilot, the new systems can summarize texts, compose emails and presentation slides, write code, and produce full essays by leveraging public information available on the web. These new systems are based on generative AI, which is the use of artificial intelligence to generate information. Although all the systems we talked about in this section use AI, from spell checkers to SMT-based translation systems, they are not based on generative AI or **GenAI**.

GenAI breaks away from trying to learn specific rules, as in the case of SMT, which tries to learn, for instance, how sentences in Portuguese map to sentences in English. GenAI systems, instead, leverage vast amounts of data and abundance of computational resources to learn "models." These models, like the ones used by ChatGPT, encode generic knowledge about grammar and language and can therefore be used for a variety of tasks. GenAI was only possible due to the vast amounts of computational power provided by cloud computing services, and large datasets publicly available on the internet.

If you were not paying close attention to this field, you may think AI was a revolution introduced by OpenAI with its ChatGPT. However, the field of AI has existed since the early days of computing. John McCarthy coined the term in 1956 at the Dartmouth College

Conference.[21] Even before ChatGPT, we have been using many AI systems daily, from Netflix and Amazon recommendations, to web search, language translation, and grammar correction.

Having said that, GenAI feels very real now and we are right to question how it will impact worker productivity. A good experiment, if there was a way to execute it, would have been to ask the Gabo of 1965 to write his famous book using all the modern technology of today. How fast could he do it? More interestingly, how would the story compare to the original one? We already have started to see some technology companies leveraging AI tools to produce very sophisticated software with a small number of employees. Cursor AI, a code editing software company, is a prime example, being very successful with only twelve employees.[22]

No one can accurately predict how people will work when these tools become even more powerful. Initial observations indicate that GenAI tools help experienced professionals, who have deeper understanding of the area and can better judge the output of the tools, more than inexperienced ones.[23] Seasoned professionals leverage these tools to complement and refine their knowledge,

21. https://home.dartmouth.edu/about/artificial-intelligence-ai-coined-dartmouth (Last accessed in March 2025).
22. https://intenvo.com/cursor-ai-user-and-revenue-statistics/ (Last accessed in March 2025).
23. https://www.nytimes.com/2025/05/25/business/amazon-ai-coders.html (Last accessed in June 2025).

rather than completely delegating the execution of tasks. We are still at early stages of the use of GenAI in companies, and we must use our human agency to make sure it will be enriching to the human experience. We'll revisit this point in the next chapter, after we go over AI and GenAI in more detail. We'll cover the details of the technology and some of its most relevant applications.

6.

AI

Context, embeddings, old school AI, hallucinations, GPT-3, system 1 and system 2, parrots, and humans

"OK, so what am I doing? Oh, I'm chasing this guy. No... he's chasing me."

Leonard Shelby in Christopher
Nolan's *Memento*

Leonard kills Teddy in an abandoned parking lot. In the back of Teddy's polaroid's picture Leonard had clearly written: "he is the one. Kill him." Since his accident, he had to rely on pictures, notes, and even tattoos to orient himself. He has anterograde amnesia. Although he clearly remembers everything before the accident, he now cannot form new memories and has short term memory loss. He knows he is Leonard Shelby, a former insurance agent from San Francisco. He also remembers his wife, who is now dead. But, unable to form new memories, he had to create a } system with tattoos in his body for the important facts, and pictures with notes to orient himself. He organized his system because he has a clear goal. To kill the person who raped and murdered his wife and caused his memory problems.

Leonard is a fictional character played by Guy Pearce in Christopher Nolan's 2000 film *Memento*. Anterograde amnesia also afflicted Drew Barrymore's character Lucy in the romantic comedy *50 First Dates*. This movie was a bit hard for me to watch, as I don't like when anything bad happens to Drew Barrymore. But at least it has a happy ending. Henry, played by Adam Sandler, falls in love with Lucy, despite the fact that every night she forgets who he is. Henry prepares a tape "Good morning Lucy" that she watches as she wakes up every day, before moving on with their lives. And they manage to live happily ever after. The case of Leonard Shelby was a bit more complicated, as he was trying to conduct a complex investigation, which by definition requires new facts and new connections, while not being able to form new memories.

Memento scenes are short and happen in reverse chronological order. This makes the audience also experience Leornard's condition. In one of the movie's funniest scenes, Leonard is running without knowing why, when he spots another person also running. He then initially thinks he is chasing the guy, just to realize the opposite is happening. He is trying to escape. By the end of the movie, we realize that Leornard has been deceived to kill others and hurt people not related to his revenge. Worse, he tricked himself into killing Teddy. It is really hard to operate with no memory. And in the case of my cherished Drew Barrymore, although "Good morning Lucy" is a cute idea for a movie with a happy ending, for someone to live a complete life, this movie would have to be longer and longer overtime. And it would have to be replayed over and over.

Let's consider that someone, or some computer program, can only keep the last five words, forgetting completely what was written before that. When reading the sentence, "Magic was certainly one of the best basketball players of all time," which has twelve words, we would only retain the last five "basketball players of all time." The computer program would not know for sure what the sentence was about. It could be about the worst players of all time. It could also be about the U.S. olympic dream team, formed with several of the best players of all time. In some unfortunate circumstances it could be about someone else, most likely not as good as Magic Johnson. The number of words we can remember in memory is our **context**. We talked earlier about the first spell checkers, which could not automatically correct the sentence "a new from of thinking" to a "a new form of thinking" because they had no context. It is like their working memory was always empty, and they just operated at a single word at a time.

Today's systems are far more powerful. Correcting "from" to "form" has become a simple task. I asked ChatGPT the following: "if $1 + 1 = 2$, $2 + 2 = 8$, $3 + 3 = 18$, how much $4 + 4$ would be?" and it gave me the correct answer, with an explanation. It calculated that $1 + 1 = 2$ was the expected case. But it also computed that $2 + 2$, which would normally be 4, was now defined as 8. Similarly, $3 + 3$ was also redefined from 6 to 18. ChatGPT told me that this was a "pattern puzzle rather than normal arithmetic" and a possible pattern was:

$(1 + 1) \times 1 = 1$
$(2 + 2) \times 2 = 8$
$(3 + 3) \times 3 = 18$

And therefore:

$(4 + 4) \times 4 = 32$

This matches exactly what I was thinking when I typed my prompt. This may seem like magic, especially when I hope that I successfully convinced you in Chapter 1 that all that computers know to do is to compute functions involving integers. There is a lot going on here, and although it is quite impressive, it is brought to you by lots and lots of integer functions. We are fortunate that computers don't mind doing all these calculations. They do it tirelessly and with very good accuracy, which is itself some form of intelligence. Before we try to unpack all of that, let's try to understand how ChatGPT solves the simple arithmetics, like *4 + 4 = 8*.

This is exactly what computers do. They compute integer functions. Therefore, this straightforward computation can be resolved with a very simple computer program:

```
int Sum(int a, int b)
{
return a + b;
}
```

This is the most efficient way to solve this problem. Just call function *Sum(4, 4)* and you'll get the correct output 8 in a single instruction. When we ask ChatGPT *"4 + 4 =,"* this is not what happens, although, after a while, it correctly answers *"4 + 4 = 8."* To understand how ChatGPT solves arithmetic problems, we need to go over a few fundamental concepts. We've already seen that context is key. Lucy cannot go through her day without watching "Good morning Lucy" and ChatGPT wouldn't be able to come up with the right answer for *"4 + 4 ="* without looking at the whole expression. In addition to having context, we need to assign some form of meaning to each word. Let's start by understanding how GenAI systems do that.

How related are queens and kings?

In the early days of Garcia Marquez's Macondo, "the world was so recent that many things lacked names." In a world with no language we may be able to have sensorial experiences but we cannot develop any form of intelligence. We need language to express concepts. Think about a simple idea. Specifying the size of objects. A primitive language, in which we have words for small, medium, and large, is much better than no language. However, if we want to be more precise, we need more dimensions. Being able to specify size by talking about length, width, and height would allow us to be more precise. If we had numbers in addition to small, medium, and large, we'd be able to be even more precise.

Now think about expressing the size of a more complex object. Figure 13 shows the measurement chart for ordering a men's tuxedo. In order for your tailor to do a good job and build you that perfect tux, you need to fill a chart with a lot of sophisticated concepts, such as "crown to cuff" and "outseam."

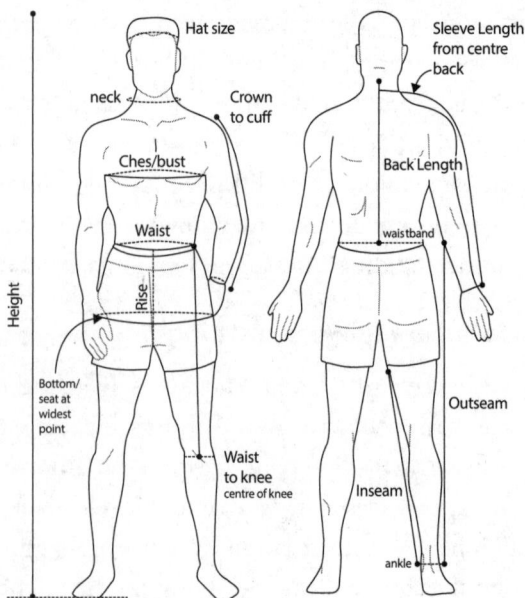

Figure 13. A typical tuxedo measurement chart downloaded from the web.

It is natural for us to think about a concept like an object's size in terms of dimensions, as we can easily map objects to our three-dimensional space. But we can also think of words as having "dimensions." The word "queen" may have values in many dimensions. Demographic dimensions could include gender, age, and income. Other dimensions may be related to their position as head of state, such as conservative or liberal. Depending on the context, "queen" could also be associated with entirely different dimensions related to rock bands, such as progressive and eighties, and in yet another context, it could be related to chess pieces, such as black or white, and movement patterns on the board.

Early computer systems that processed language, like spell checkers, web search, search advertising, and machine translation, didn't represent words as a set of dimensions. In web search, we store the information of which words are present in a web page. For each word, we may additionally store their occurrences within the page and some salience information, indicating if the occurrence happened on the section title, or if it was in bold or italics. All this information goes into the inverted index, as we discussed in Chapter 3. However, if the word "queen" appears in a given page, we don't normally use its different dimensions, such as progressive rock band, valuable chess piece, or head of state. Search doesn't use context either. That is why it is still possible for us to see printer advertisements as results to query "latin cannon" in some web search engines.

When we are trying to write, or generate text using GenAI tools, a deeper understanding of words and their context is essential. Just to be able to visualize these concepts, let's go back to object sizes. We may associate "elephant" and "skyscraper" with "big" and "ant" and "tent" with "small," as illustrated in Figure 14. Writers that understand these associations can compose sentences such as "For her, that elephant was as gigantic as the Empire State building."

For that reason, Yoshua Bengio and his collaborators came up with the concept of word **embeddings**.[1] Think of embeddings as a vector representation for each word, with one value per dimension. In our example, we could have the embeddings shown in Figure 14.

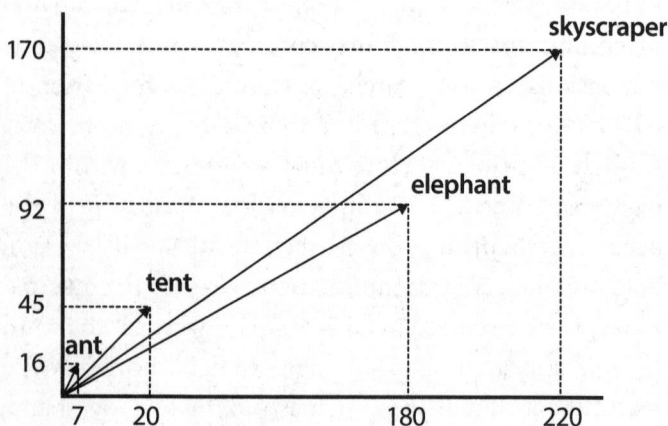

Figure 14. A few embeddings. *E(word)* is the embedding vector for "word." In this example,
E(elephant) is *(180, 92)*, *E(skyscraper)* is *(220, 170)*, *E(ant)* is *(7, 16)*, and *E(tent)* is *(20, 45)*.

---

1. A Neural Probabilistic Language Model, Y Bengio and others, Journal of Machine Learning Research, March 2003, 1137–1155.

As shown in the figure, the vectors for elephant and skyscraper are much closer together than the ones for elephant and ant, in the size dimension. In reality, there are way more dimensions than just size. In the "animalness" dimension elephants and ants would be very related. Think of embeddings, as mapping each word to a set of dimensions that captures both the semantics, what the words mean, and syntax, how they are used in sentences. Animalness and size capture semantics, while being a verb or an adjective capture syntax. Some dimensions may capture both a bit of syntax and semantics, such as being feminine or plural.

As an example, let's consider that our vocabulary has one thousand words and one hundred dimensions. Each word would be then represented as a vector of one hundred positions and we'd represent embeddings as a matrix of dimensions one thousand by one hundred, as shown in Figure 15. In this representation, each word in the vocabulary is mapped to an index number, for instance, $a = 1$, *aardvark = 2*, and so one, until *zebra = 1000*. The same is true for dimensions. We could have *size = 1, animalness = 2, verb = 3*, and all the other dimensions until 100. The cell *[2, 3]* in the matrix would be the value for aardvark's verb dimension, which would likely be very low as it is a noun and not a verb.

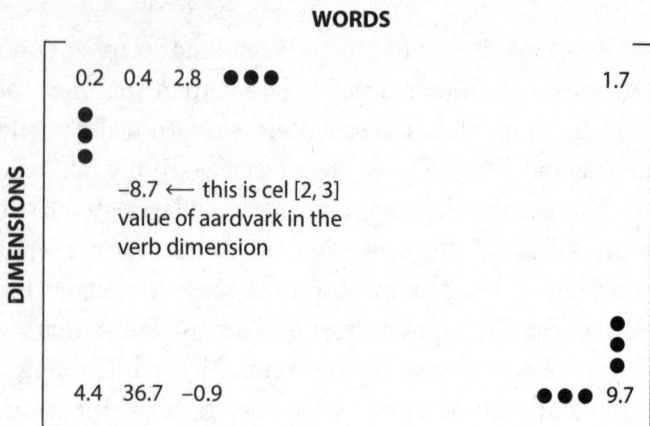

WORDS

| 0.2 | 0.4 | 2.8 | ● ● ● | | 1.7 |

−8.7 ← this is cel [2, 3] value of aardvark in the verb dimension

4.4 36.7 −0.9   ● ● ● 9.7

Figure 15. Embedding matrix in which the columns represent the embedding vectors for all words in the vocabulary. The size of the matrix is *1000 × 100*, as we have one thousand words and one hundred dimensions.

In this representation, words are vectors in a **high-dimensional space**. A space of exactly one hundred dimensions. And similar words would have similar vectors. We'd expect that the vectors for "queen" and "king" are very close together, while the vectors for "queen" and "aardvark" shouldn't be very related. The way we compute the similarity between two vectors is by their **dot product**. The dot product between two similar concepts is positive, while it is 0 for unrelated concepts, and negative for opposing concepts. Figure 16 shows an example in two dimensions.

What we observe in the models used by ChatGPT and Microsoft Copilot is that not only the embedding vectors for "queen" and "king" are related. As shown in Figure 17, their difference is also similar to the difference between the embeddings of "woman" and "man." We can write the equation:

E(queen) - E(king) = E(woman) - E(man)

As we can see in Figure 17, these differences are vectors pointing in the same direction. It is as if the high-dimensional space of embeddings assigns a direction for gender.[2] I extracted this example from an online video that does a great job describing these concepts visually.[3] In fact, as the video explains, "queen" is not exactly the feminine of "king," as it could also be the rock band, and either "king" or "queen" could also be referring to chess pieces. For that reason, the relationship between the embeddings of "aunt" and "uncle" or "niece" and "nephew" would be even more similar to "woman" and "man" than "queen" and "king:"

E(woman) - E(man) = E(aunt) - E(uncle) =
E(niece) - E(nephew)

---

2. Please see a visualization of embeddings at https://digitalagencybook.org/visualizations/word-embeddings.
3. https://www.youtube.com/watch?v=wjZofJX0v4M&t (This is a great video, and the 3Blue1Brown channel has several videos that use animation to better explain complex problems. Last accessed in April 2025).

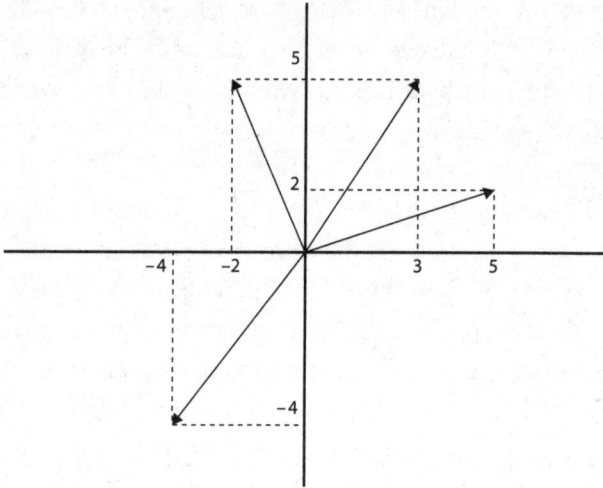

Figure 16. Dot product examples. The dot product of *(3, 5)* and *(5, 2)* is *3 × 5 + 5 × 2 = 25* (correlated, vectors in the same direction). The dot product of *(5, 2)* and *(-2, 5)* is *5 × -2 + 2 × 5 = 0* (unrelated, vectors are orthogonal). The dot product of *(5, 2)* and *(-4, -4)* is *5 × -4 + 2 × -4 = -28* (opposing, vectors are in opposing directions).

There are more relationships that hold true. For instance, *E(men) - E(man)* should capture plurality, such that, if we want to find out the embedding of "grapes," we could approximate it by:

$$E(grapes) = E(grape) + E(men) - E(man)$$

Embeddings are quite powerful, as these examples illustrate. All GenAI systems use embeddings as their starting point. The notion of embeddings is so fundamental that, in 2018, Bengio was one of the recipients of the Turing Award, the Nobel Prize for computer science, given his fundamental contributions to AI. Bengio is also the most cited computer scientist, both in total number of citations and h-index. His current h-index is 247, meaning he has 247 papers with more than 247 citations each. He is also the living scientist, across all fields, with the highest number of citations and was included in the 2024 TIME Magazine list of most influential people. An AI rock star!

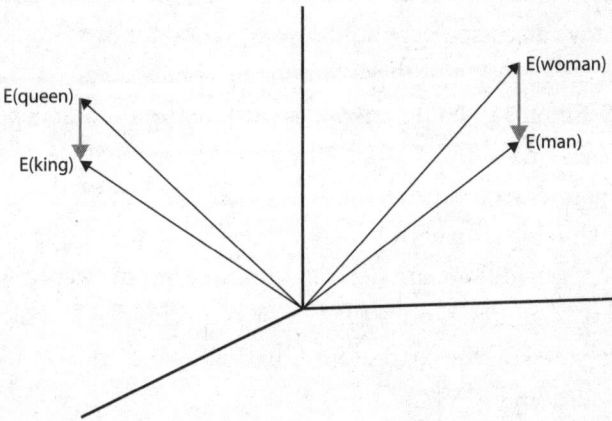

Figure 17. This is a 3D projection of the high-dimensional space of embeddings. The difference between the embeddings of "queen" and "king" is similar to the difference to the embeddings of "women" and "men."

There are two important points about embeddings we need to clarify further. The first is that we don't compute embeddings for words but for **tokens**. In the same way that in web search we added concepts, such as "bubble sort" to the inverted index, a token can be a combination of words, like "bubble sort" or "United States," a simple word, such as "queen," or subwords, like the prefix "un" or the suffix "ing."

The second point of clarification is that, in reality, dimensions don't cleanly map to single concepts like "size" or "animalness." GPT-3, one of the models ChatGPT uses, has 12,288 dimensions. Although all the properties that we discussed here hold true, it is hard for us to understand what each of these dimensions really mean. A single dimension could encode a combination of concepts. This is one of the reasons why we often say that we don't fully understand these models. Making AI models more explainable is an active area of research. Before exploring how embeddings are derived and used more deeply, we need to understand a bit better old school AI, which was state-of-the-art before ChatGPT and the GenAI revolution.

One important side note. We must distinguish the application, ChatGPT or Copilot, from the model, GPT-3 or Gemini 2.5. GenAI models are generic and can be integrated in many applications. Copilot is part of the Microsoft productivity suite and uses AI models to help you with your Excel spreadsheets and Powerpoint presentations. The same models can be used by ChatGPT for prompts and by software development services for writing code. Most of this chapter is dedicated to explain how the models work.

Old school AI

Before AI was generative, it was predictive. In a gross generalization, AI systems used a bunch of historical data to learn patterns. Based on these patterns, it could make predictions about future behavior. By looking at historical sales data, we can predict how much we'll sell in the future. In the same way, by having users watch and label cat videos, an AI will be able to correctly predict if a new, unseen video, is feline-related with reasonable accuracy. That is one of the reasons the ESP game, which we discussed in Chapter 2, was so interesting. It produced labels to **train** AI systems.

Training means going over historical data, or data produced by users, and "learning" from it. Before GenAI, learning was targeted to a specific task. We could, for example, train a model to be able to answer how much we'd sell in the future. The parallel to human intelligence is key here. Someone with extreme dyscalculia would have to go through many examples of arithmetic problems to learn how to answer 7×8 on their own. A math genius, on the other extreme, may be able to understand how multiplication works with very few examples. When looking at cat videos, efficient training algorithms will learn the discerning **features** that separate the videos labeled as cat from the non-cat videos. A feature for identifying cats can be their jewel-toned eyes, often green, yellow, or blue. Good features are good predictors. The AI model we want to train is a compact representation of these predictive features that can, for instance, be used to output "cat" or "no cat" when prompted with a new video it has never seen before.

An example of a simple predictive task is to estimate growth of a given service. Figure 18(a) shows an example. Since the points seem to fit in a straight line, we can use **linear regression**, which is a model that computes the lines that best fits the historical data. In this case, we just need to store the two parameters for the line, slope and y-intercept, to be able to extrapolate the data and make predictions about the future. Of course, growth in the future may be very different from this, if past data proves to be a poor predictor. The reasons may be an upcoming recession, or a latent problem with the service that turned new potential users away. Identifying when data is not suitable for a given task may be difficult for both AI or humans. Sometimes, new data sources can help. For instance, we could try to correlate the historical growth data with economic trends, to avoid mispredictions when recessions happen.

A more complex problem may be identifying cat videos. Let's drastically simplify it for the sake of this presentation, and consider that there are only two discerning features. Their distinctive eye features and the fact that cats have whiskers. If we have a model that understands the pixel patterns for these two features, we produce a two-dimensional vector <has cat eyes, has whiskers> for each video. The values in these vectors represent probabilities. For instance <0.1, 0.9> means that we likely didn't detect cat eyes in the video but we've detected whiskers with 90% of probability. 15(b) shows, for every video in our **training set**, which are the videos that have been labeled by users as "cat" or "no cat," where they land in this two-dimensional space. In this scenario, there is a line that clearly separates the cat-labeled videos from the others. Crunching through the data points and finding the line that best separates the videos is the job of our **training algorithm**. When we identify the best line, its representation would then be the AI model that we'd use to classify new videos. Any video which lands above the line would be classified as a cat video, while the ones that land below the line would be classified as not related to cats.[4]

This way to solve the cat video problem is easy to describe, especially because it is easier to draw figures in two dimensions than in a high-dimensional space. Unfortunately, it is unlikely our solution would produce very good results. The reason is somewhat intuitive. A video encodes a lot of information. A typical YouTube video is a few minutes long, at 24 frames per second. Each frame has 3840 × 2160 pixels. This is a lot of

---

4. Please see a visualization of this cat video classification algorithm at https://digitalagencybook.org/visualizations/cat-video-classifier.

information, that we'd be converting down to two numbers, the probability of having cat eyes and the probability of having whiskers. This is called **dimensionality reduction**. Very aptly named, it simply means mapping from a high dimensional space into a low dimensional one. In this case, instead of looking at all the pixels of every frame to make a decision based on the entirety of the video's content, we are relying on two dimensions only. The reason that old school AI relied on a small set of dimensions, each hopefully corresponding to a feature of good predictive accuracy, is related to how we think about problems.

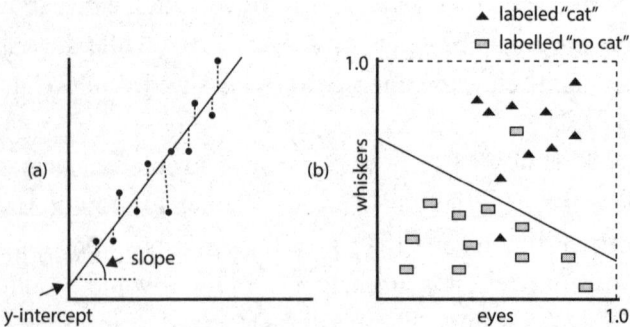

Figure 18(a). AI model to predict future growth of a service. (b) AI model to classify videos as "cat" and "no cat" using two simple features, the presence of whiskers and cat eyes.

If someone asked you what is your mental process for detecting cats in videos, you may say something like "I look for the feline features," or "I look for whiskers." In reality, we likely do it by processing the high-dimensional information as a whole, the whole image, pattern matching it to our mental model of cats. But we really don't fully understand how we do it, and therefore, we don't know how to explain it. This is why, if you pose the problem of writing a program of classifying videos as "cat" or "no cat" to a software developer, it is likely that they would come up with an algorithm that resembles the way they think. Something like, "I just look for whiskers, so I'll write a whiskers detector algorithm."

The whiskers detector program may use all the information in the video, processing all the pixels in every frame to find patterns that resemble whiskers. Implemented this way, it can be viewed as a rule-based system: if you clearly see this shape, assign high probability. If you see half of the shape, assign a lower probability. As we discussed, the first machine translation systems were also rule-based. Words were translated with predefined dictionaries and we'd have encoded rules that mapped sentences from one language to the other based on grammatical structure. The advantage of this approach is that it is very explainable. For each video, we would know why it got the values for its two dimensions, has whiskers and has cat eyes. We also can easily understand what these dimensions mean.

When I started explaining embeddings, I also pretended that their dimensions mapped to well-defined concepts, like animalness and size. In reality, it is very hard for us to understand what the dimensions in a

large language model (LLM) mean. Dealing with a small number of explainable dimensions is much easier. However, dimensionality reduction has the major downside that it leaves information on the table. That is why we moved away from old school AI and embraced high-dimensionality, at the expense of interpretability of the AI models. The GPT-3 model has a high number of dimensions, 12,288 per token. By giving up on the idea of well-understood features, such as cat eyes and whiskers, we can use a high-dimensional space to represent each video. In this new representation, dimensions have no explainable meaning. By doing this shift, from few explainable features to many opaque ones, we move away from old-school AI to modern AI. It's very likely that modern AI is more aligned to how our brains process information.

With modern AI, we'd completely redefine the training problem. Instead of simply having to find a line that separates the videos, as shown in Figure 18(b), we'd have to first compute the embeddings for each video, which would produce a high-dimensional vector per video. The embeddings would encode space and movement patterns. We'd then have to separate the videos in the same way we did in the two-dimensional (2D) space when we had only two features. In the same way a point separates a line and a line separates a 2D plane, a hyperplane separates a high-dimensional space. I loved when Han Solo took the Millennium Falcon into hyperspace in Star Wars. Hyperspace in AI is far less exciting than that. A hyperplane is just a simple way to split the embedding space to produce two clusters of videos, "cat" and "no cat." The embedding vectors for cat videos would be on one side of the plane while the non cat videos would be on the other side.

In old school AI, finding the discerning features was almost like an art. We'd have to understand the data well enough to have a good intuition of which features would be predictive or not. Selecting the correct set of features for a given task was really hard. With modern AI, we don't do any of that. After trying to find the hyperplane that separates the embedding space, we may not achieve good results. Visually, this would mean that no plane cleanly separates the data, leaving the cat videos on one side and the non-cat ones on the other. Not a problem. We'd go back to refine the embeddings, and we'd keep repeating this process until we have good embeddings that allow us to cleanly separate the cat videos from the rest.

We can even start with very simple embeddings, based on simple pixel and movement patterns. At each interaction, we would find the best hyperplane that separates the embedding vectors, and compute the error: how many videos in the training set were wrongly classified. Then we'd refine the embeddings and keep repeating the process, until we see no further improvement in the error rates. We replace the work engineers and data scientists did, sorting through data and hand picking the best features, with number crunching. Taking advantage of the availability of data and the abundance of computational resources, and the fact that computers don't mind doing the same computations over and over, we can compute the best AI model for the tasks we want to solve. Modern AI training is antifragile. With more data, the system will improve. It will recover from its own failures. It does not require talented engineers and data scientists to hand-craft features based on their domain knowledge.

With this modern AI implementation, once we find the best embeddings and the hyperplane that separates them, they would become our AI model. Using this approach, there is no way for us to easily understand why a video was identified as "cat." But the accuracy of the results is likely much better than in our original, hand crafted 2D implementation. Now that we are relieved that YouTube will never show us dog videos when we ask for cats, let's go back to how ChatGPT works.

It's all about the next word

One fact most people may have heard about GenAI is that all it does is predict the next word. This is largely true, but let's try to unpack it a bit more. We know we have the embeddings for all the tokens in the input prompt. That is our starting point. But how do systems like Microsoft Copilot produce the output for our prompts? It seems unlikely that AI systems would be able to generate complete answers to complex prompts with a model that only knows the probability of the next word in a sentence.

Let's start with a simple example. For now let's forget about embeddings, training sets, and all the AI concepts we've been talking about. Let's say we just have one of the most famous English sentences to work with: "to be or not to be."

We can calculate some probabilities. After the word "to," we always see "be." After "be," half of the time we see "or" and the other half we see a stop sign. If we continue with this line of thinking, we get to this table:

to: next word is "be" (100%)
be: next word is "or" (50%) or stop sign (50%)
or: next word is "not" (100%)
not: next word is "to" (100%)
stop sign: next is action stop writing (100%)

This is a very simplistic AI model. It only knows five tokens. The four words in the sentence "to," "be," "or," and "not" and the stop sign. This would not be a very good AI model for anything at all. But let's say we use it with the prompt "to." Using the find the next word algorithm, it would produce this sequence in order:

Input "to," next word "be" with 100% probability. Answer so far "to be"

Input "to be," next word "or" with 50% probability or stop sign with 50% probability. Since the probabilities are the same, the AI would arbitrarily choose one option from the two presented below.

Option 1. When the stop sign is chosen.
Input "to be." The next step is to stop writing. The final answer would be "to be."

Option 2. When "or" is chosen.
Input "to be or," next word "not" with 100% probability. Answer so far "to be or not"
Input "to be or not," next word "to" with 100% probability. Answer so far "to be or not to"
Input "to be or not to," next word "be" with 100% probability. Answer so far "to be or not to be"
Input "to be or not to be," here again we could fork into the same two options. We'd either add a stop sign and stop writing, or we'd continue following the same pattern.

If we put these two options together, our AI would produce one of the following answers to prompt "to:"

to be.
to be or not to be.
to be or not to be to be.
to be or not to be to be or not to be.
(and so on)

We don't know which of these answers would come out. Even without any sophisticated training, our AI learned to write a few patterns, like "or not to be," based on the single sentence it was trained with. This is a very silly example in many ways. If the AI system just learns one sentence, it would very likely be able to reproduce that sentence even if it is not very intelligent. The only reason we get some variation here is because the word "be" appears twice, followed by either "or" or the stop sign.

Models like GPT-3 are trained with a lot more than one single sentence. They use large parts of the web, including all of Wikipedia, books, and other publicly available datasets. In the next section we'll go into the main components of the GPT-3 model. But first, let's review how the model is used by applications such as ChatGPT and Microsoft Copilot. Independently of the application, all the models work the same way. Computing the best next word.

ChatGPT combines any directions you may provide, like which traits it should have, with your prompt to generate the input. These directions are sometimes called the **system prompt**. The output it produces is a continuation of this combined input, just as in the "to be or not to be" example. As we'll see next, the embeddings for each token in the combined input are used to derive the next tokens that should be added to the input, one by one, until the full output is produced. For now, let's just assume that we have a current sequence of tokens, consisting of any directions in the system prompt, the actual prompt, and the output produced so far. The **inferencing** algorithm will then produce a list of possible next tokens with accompanying probabilities, as shown in Figure 19.

While **training** produces the AI model from available datasets, the training data, **inferencing** uses the already pre-computed models to answer queries.

In the case of cat videos, inferencing was using the model to classify a new video as "cat" or "no cat." For generating text, inferencing is using the model to compute the next words. The inferencing algorithm will select which word to add based on its probability. In the example of Figure 19, "Drew" would be selected with 6.7% probability, "Barrymore" with 6.4%, and so on. Some models may allow you to set a **temperature**. A high temperature allows the inferencing algorithm to choose tokens with lower probabilities. Setting the temperature very high will produce very creative, and more inaccurate, outputs.

Besides the fact that the only thing GenAI models know how to do is computing the probability of the next word, the other popular tale about them is that they **hallucinate**. This is also true. There is nothing in training or inferencing that protects AI models from producing a completely random response. This is particularly true if we set the temperature high. It is sometimes surprising how these models produce amazing responses for certain prompts, while completely failing in others. The reason is that, while in old school AI we had explainable features and well-defined rules, GenAI models are completely based on the embeddings and their associations, learned from the training data.

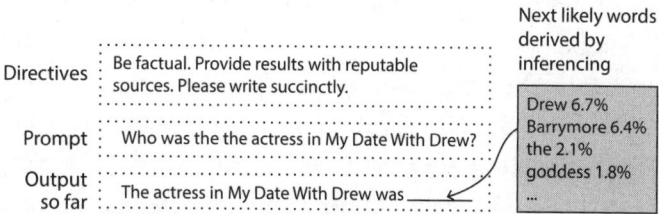

Figure 19. The inferencing algorithm produces a list of next likely tokens that will be appended to the output, one by one, until the full output is generated. After each token is selected, we invoke a new round of inferencing, producing a new list of next likely tokens.

All of us who use GenAI tools regularly have experienced these hallucinations. A funny one that happened to a friend of mine went like this:

GhatGPT: "Got it! Just call my name when you need me."
My friend: "By the way, what's your name again?"
ChatGPT: "You can call me **Siri**. Feel free to reach out anytime!"
My friend: "Like Apple's Siri or something else?"
ChatGPT: "Oh, my mistake! I'm actually ChatGPT. You can call me that."

If we think about AI as an intelligent being, it is quite surprising it would be confused about its own name. If, on the other hand, we think about AI for what it actually is, a system that learns from publicly available texts and is able to generate text based on a model that predicts the likelihood of the next word, we can better understand this scenario. There are many texts on the web that say "you can call me Siri." Therefore, the computed likelihood of "Siri" appearing after "you can call me" is probably very high. There is nothing on the model that would prevent mistakes like this. GenAI models don't know how to distinguish facts from falsehoods. They only understand tokens, and there are no good and bad tokens. We could, of course, add hard coded rules to process the generated text and block a few embarrassing blunts, but we'd need a lot of rules and it would be impossible to prevent all of them.

In traditional, old-school AI systems, model success hinged on two difficult tasks. Collecting the right data and manually identifying the right features. Engineers needed to deeply understand the domain to extract human-interpretable features, like "has whiskers" or "cat eyes." We assumed these features to be predictive. Even if the dataset was complete, a poor feature set would cripple model performance. In contrast, modern generative AI systems like GPT-3 shift the burden of selecting appropriate features away from the hands of software developers and data scientists. They automate feature learning through large-scale embedding models, learning high-dimensional representations directly from raw data. This relieves the need for handcrafted features, but places even greater importance on curating high-quality, diverse, and task-aligned training data. If foundational data is biased, irrelevant, or lacks coverage of the inferencing tasks, the learned embeddings will be distorted, and the produced outputs will be unreliable, regardless of the model and the size of the training set.

Unlike traditional systems like PageRank, which explicitly rank pages based on citation-like link structure, LLMs do not have built-in schemas for source credibility. Instead, they passively absorb statistical patterns from their large datasets, which may include a mix of reputable, neutral, and dubious sources. During training, if a peer-reviewed journal appears consistently in trusted contexts and is referenced across high-quality documents, the model may implicitly assign higher likelihood to its pattern. But this is an emergent behavior, not guaranteed. There is no built-in epistemology. Engineering credibility into LLMs requires intervention, such as curating the data used in training. Without safeguards, LLMs are vulnerable to hallucinations and misinformation poisoning.

Now that we have a good understanding about embeddings, training, and inferencing, we have a solid foundation to understand the inner workings of GenAI models a bit better. Next we'll go over the main aspects of how GPT-3 works under the covers. Starting from the embeddings, what are the steps it takes to produce the list of likely next tokens that are added to the output.

Attention is all you need

Let's start by defining what GPT-3 means. G stands for generative, which is easy enough to understand. Afterall, our objective is to generate text. P stands for pre-trained. That is, in my opinion, the most important feature of these models. It means that GPT-3 was already trained to understand language and generate text. The reason this is so important is that training these large-scale models is very expensive, consuming a lot of computational resources for a considerable amount of time. It is estimated that GPT-3 takes about one week to train.[5]

If we want to generate a marketing campaign for our company, the fact that GPT-3 is already pre-trained means we don't need to teach it to produce marketing materials. It very likely already learned that from publicly available texts. We only need to provide the information about our company as part of the prompt, and it will likely do a great job in producing the marketing campaign, being consistent with the information we provided as input. This consistency comes from the last letter in the acronym, T, which stands for transformers.

Transformers are an AI technique introduced in the 2017 paper, *Attention is all you need*.[6] It is the foundation for all the GenAI models we have today. There are two

---

5. The Cost of Training NLP Models: A Concise Overview, O. Sharir, B. Peleg, and Y. Shoham, available at https://arxiv.org/abs/2004.08900 (Last accessed in April 2025).

6. Attention is All you Need, A. Vaswani, N. Shazeer, N. Parmar. J. Uszkoreit, L. Jones, A. Gomez, L. Kaiser, P. Łukasz, and I Polosukhin, 31st Conference on Neural Information Processing Systems (NIPS). Advances in Neural Information Processing Systems. Vol. 30.

main reasons transformers became so popular. The first, comes from the title of the paper: **attention**. It means that transformers can focus on the salient parts of the text. Imagine being at a party, with a lot of background noise. You're not able to pay attention to every conversation, but you can pay attention to important conversations around you, and even switch focus once you hear your name. The same is true for the attention mechanism in the transformer algorithm. It can distinguish the relevant parts of the input text from the background noise.

The second aspect of transformers that made the technique so popular was its efficiency. While previous approaches required processing the input text sequentially, one token at a time, transformers allow us to process it in parallel. All tokens at the same time. This allows us to drastically reduce the time to process the input text, enabling us to use larger and more complex models. This is also why GenAI applications, such as Microsoft Copilot, can answer complex prompts so fast and the reason we use GPUs, graphical processing units, for GenAI computations. GPUs provide a larger number of specialized processor cores that can be used to run the transformer computations in parallel. We have already seen that the answers are generated by adding one token at a time to the output. But what are the exact steps that happen when we pose a prompt to GPT-3?

As we mentioned before, we start with the embeddings for each token in the input. Since each token has an embedding vector, the starting point is just a collection of vectors. Initially, each of these vectors are the original embeddings for each token. These initial vectors were computed during training, and at this point don't have any information derived from the user prompt. We then start manipulating these vectors so they will have more and more information from the context provided in the input. For instance, the token "magic" would likely have, in its original embedding vector, high weight values in the dimensions related to tricks and magicians. However, if the input is something like "what are the highlights of the basketball player Magic Johnson's career," as we process these vectors we'll reduce the weights in the sorcery dimensions and increase the weights in its basketball dimensions.

The way we manipulate these vectors is using matrix multiplications. If we think about the values in the embedding vectors as weights, when we multiply matrices we are increasing or decreasing these weights. One of the main characteristics of LLMs is the **number of parameters**. This simply means the number of weights in the several matrices it uses. All these matrices are computed during training. The first of them is the embeddings, which in GPT-3 has a total of 617,173,056 parameters:

Size of GPT-3 embeddings = 50,257 tokens
× 12,228 dimensions = 617,173,056 parameters

The embeddings represent only 0.35% of the total 175 billion parameters in GPT-3. All the other parameters belong to several other matrices, which are used to alter the weights of the original vectors during inferencing, as in the Magic Johnson example we just discussed. If we squint, there are two main manipulations we do to the input vectors. The first is **self-attention** and the second is **feedforward neural network**. Let's talk about each one.

In the self-attention step, we process every token in the input and partial output generated so far, considering the contributions from surrounding tokens. GPT-3 uses a **context size** of 2,048 tokens. This means that every token can have its weights modified by the 2,047 tokens that precede it. This is true not only for the original input, but also for every new token we add to the output. The context size does not impact the number of model parameters, but it does impact the processing time and the computational resources used in the self-attention step. If we again consider our example sentence "what are the highlights of the basketball player Magic Johnson's career," the self-attention step will identify that "highlights" refer to "Magic Johnshon's career" even though there are other tokens, "of," "the," and "basketball," in the middle of the sentence. As we discussed before, the embedding dimensions will capture both semantic information, such as the fact that "magic" here refers to the player and not to sorcery, and grammatical information, such as that the highlights of Magic's career are the subject of the question.

You can think of the self-attention phase as modifying the dimension weights of each token based on the context. Tokens that are not in the context window do not come to play during the self-attention phase. For instance, although Magic Johnson is associated with the Los Angeles Lakers, since Lakers doesn't appear in the context, this association will not be captured during self-attention. That is when the feedforward neural network phase comes to the rescue. Feedforward happens after self-attention. And like self-attention, it also works by modifying weights in vectors representing tokens. But this time around, the goal is to capture extra connections that were not present in the input, like the fact that Magic played for the Lakers.

As with embeddings and self-attention, feedforward also relies on matrices computed during training. There are many occurrences on the web and in the data sources used to train models that indicate that Magic played for the Lakers, and that his career highlights included being five times champion of the National Basketball Association (NBA) and three times most-valuable player (MVP). These relationships are somehow encoded in the matrices used by feedforward. During inferencing, they are used to add more context to each token. For instance, once in the self-attention phase we identify that the token "highlights" refers to Magic's career, the feedforward phase will be able to connect it to being elected NBA MVP three times. It's like feedforward brings in the new tokens that may be part of the response.

We call a self-attention step followed by a feedforward step a **transformer block**. In GPT-3, we execute 96 transformer blocks for each token. This means that, after constructing the inputs by coalescing the embedding vectors for all the tokens in the system prompt and user prompt, there are 96 steps of attention and feedforward, one after the other. These steps will modify the vector weights to capture all the necessary relationships so that we can select the next token. This is a lot of matrix multiplications. But by now you already know that computers don't get tired and don't complain about multiplying weights over and over.

Each large language model has a different number of parameters and transformer blocks. In GPT-3, we've already seen that only 0.35% of the parameters encode embeddings. Most parameters go to the transformer blocks. Although the title of the paper is *Attention is all you need*, only about 33% of the 175B parameters are used for self-attention. Feedforward uses the most parameters, about 66%. This intuitively makes sense, as feedforward is where most associations are stored.

After we are done with the 96 transformer blocks, we are still not done. At this point, the vectors have been modified to capture all the information the model has about each token in the input. But we still need to select the next token. As we are not very creative, we'll again use the same technique used everywhere else. Matrix multiplication. We multiply the vector of the last token in the input by a **de-embedding matrix**, which maps the embedding dimensions back to the token space. It is the transpose of the embedding matrix. Instead of mapping tokens to dimensions, it maps dimensions back to tokens.

In the case of GPT-3, the embedding matrix has size 50,257 tokens by 12,228 dimensions, while de-embedding has size *12,228 × 50,257*. By multiplying the vector of the last token in the input, which has size 12,228, by the de-embedding matrix, we get a vector of size 50,257, which is the size of our vocabulary. The weights of this vector will represent the probabilities for the next token. If we go back again to our example "what are the highlights of the basketball player Magic Johnson's career," and we haven't yet added any token to the output, the last token would be "career." At this point, the vector for "career" has been modified by the 96 transformer blocks we discussed above. We'd then multiply it by the de-embedding matrix. The output would be a vector of probabilities, for instance:

Magic: 5.2
Highlights: 5.1%
NBA: 4.9%
MVP: 4.7%
Lakers: 4.6%
Los: 4.4%
Angeles: 3.7%
...

Although I showed only tokens with the highest probabilities in this example, the output vector would contain probabilities for every token in the vocabulary. We'd then select one of these tokens, likely "Magic" or "Highlights," add it to the output, and run everything again. Don't worry, as I mentioned before, computers don't get tired and don't complain. And this is basically it. These are the key steps all GenAI systems use.[7] The only major aspect of the system I left out is user feedback mechanisms. GPT-3 uses **reinforcement learning from human feedback (RLHF)**. This sounds complicated, but it basically means that humans can provide feedback, thumbs up or down for instance, which is then used to refine the models. We can also use human judgments to rank how well different versions of the model solve the same tasks. RLHF is not novel, we have been using it for years for many tasks, such as web search, as we discussed in Chapter 3. A very interesting documentary on using reinforcement learning for gaming is *AlphaGo*.[8]

If someone explained the inner workings of GPT-3 to me before I had hands-on experience with the system, I'd never have guessed it would work so well. It is fascinating how much information we can encode in 175 billion weights, split into encodings, self-attention, and feedforward matrices. Even the engineers working on the project were impressed that it worked so well for such

---

7. Please see a visualization of the inferencing algorithm for GPT-3 at https://digitalagencybook.org/visualizations/gpt3-inference.
8. https://en.wikipedia.org/wiki/AlphaGo_(film) (Last accessed in April 2025).

varied tasks. I'd never have guessed that it would be able to produce sound text without any explicit knowledge about grammar. There are no grammar rules in the system. All is done by computing the next best word. And we are still in the early stages of AI. These models will continue to evolve, and new techniques that will deem transformer blocks obsolete may be invented. But for now, transformers are here to stay and most of the innovation has been on increasing the number of parameters. Let's go over how the size of the model affects the quality of the output in a bit more detail.

Bigger is better, at least for now

Now that we understand how the system works for generating text, let's go back to the problem of how it can be used for math. Computers are very good at solving integer functions. In fact, that is the only thing they can do. That is the reason we map even tasks that, at a first glance, have nothing to do with integer functions, such as generating text, into algorithms that, at their core, just do a bunch of matrix multiplications. Solving $x + y$ is easy, and can be done efficiently by a single computer instruction. Just plug in *4 + 4* in your calculator and you get 8 back right away.

When we enter *"4 + 4 ="* as the prompt for any GenAI system, we also get the correct answer back. However, that is done using the same inferencing algorithm we just discussed. The prompt is divided into tokens. We get the embeddings for these tokens, run the many layers of transformer blocks, and finally reach the conclusion that "8" is likely the most probable token. The self-attention block will identify that the equality sign refers to the sum *"4 + 4."* We get the right answer because we have a lot of texts on the web that have the sentence *"4 + 4 = 8,"* so we can learn this association during the feedforward phase. Needless to say, this ordeal takes much more than a single instruction.

GPT-2 was much smaller than GPT-3. It had only 1.5 billion parameters for its largest model. Although GPT-2 was impressive for many tasks, it didn't do a great job for math problems. It would likely be able to solve *4 + 4 = 8*, but it quickly stopped working when the problems got a little bit more complex. The reason being that, with a much smaller number of parameters, we are only able to capture a smaller number of concepts and associations. In addition to the number of parameters, GPT-2 was trained using a much smaller dataset than GPT-3. The training set for GPT-3 had 300 billion tokens while GPT-4 had 1.8 trillion tokens. With less training data, we have less opportunity to learn new concepts. In fact, with GPT-2 we would be better off asking the system to generate a computer program that solved the math problem, rather than directly asking for the solution. Perhaps GPT-2 had an easier time "learning" programming languages than more general math concepts. That may be because programming languages are more similar to natural languages than math, or because GPT-2 had more code than math in its training set.

The focus on bigger models is justified. By using more training data, more parameters, and larger context sizes, we can solve harder prompts more accurately. For a human to read all the training data used by GPT-3, it would take 30,000 years of continuous reading, eight hours a day at 250 words a minute. Even if a human could read it all, it is unlikely that it would be able to retain most of it. But we know that with 175 billion parameters, GPT-3 can retain a lot of information. In a way, we are just throwing money at the problem. Larger models are more costly for both training and inferencing. This means more computational resources, including more energy. If we continue betting on size, the limiting factors for progress in AI will be access to more data and computational resources, especially hardware and energy. Another downside of betting on larger and larger models is that only a few countries and a few companies will be able to compete. Even universities may be left out of the race to advance the state-of-the-art in AI.

Although the number of parameters of GPT-4 is undisclosed, we estimate it is in the order of one trillion. Being much bigger allows it to do some very impressive things. As we discussed in the beginning of this chapter, it is able to correctly answer "if *1 + 1 = 2, 2 + 2 = 8, 3 + 3 = 18*, how much *4 + 4* would be?" Now that we know a lot more about how GenAI models work, we can appreciate how complex this task is. There is no way that we can type this statement in a calculator and expect to get the right answer. Also, there is no way to write a computer program to solve math puzzles like this one in a generic way. The best a programmer would be able to do is to try a few heuristics that would be able to solve simple patterns. GPT-4 solves it, and it feels like magic.

Somehow it is able to identify that we are posing a math puzzle. This is likely done by noticing the inconsistency on what is before and after the equal signs. With enough training data showing examples of several math puzzles, these larger models can capture the transformations normally used to solve these puzzles and encode them as associations in the model matrices. When I asked the same puzzle, but with slightly higher numbers, GPT-4 could still solve it, but GPT-4 Mini, which is a smaller model, gave me the wrong answer. For prompt "if 17 + 17 = 461, 18 + 18 = 648, 19 + 19 = 722, how much 20 + 20 would be?" it wrongly answered 806 instead of 800. The final paragraph of the answer was: "It looks like the relationship between the numbers is more complex, possibly involving a different operation or even pattern within the digits. However, without a clear and simple arithmetic pattern emerging from the numbers given, a reasonable answer for 20 + 20 following a similar growth trend would likely be 806 based on the continuing pattern."

This emergent mathematical reasoning in GPT-4 was not present in the previous models. The much larger training set, including math textbooks and online educational websites, such as Khan Academy, alongside the much higher parameter count have helped. GPT-4 can now solve difficult problems, such as the sum of the first one hundred integers, just because it has seen it many times in textbooks during training and could capture it in its parameters. It does not use a calculator, and relies only on the transformer blocks we discussed before. However, it also introduced two new concepts that also helped with its math solving skills: **chain-of-thought (CoT) reasoning** and **tool integration**. We'll go over these soon, but first I'd like to step back and examine how humans think.

System 1 and system 2

Although Will Smith initially appeared to laugh at the joke, his wife, Jada Pinkett Smith, was clearly uncomfortable. Moments later, he rose from his seat and slapped Chris Rock on the face. He then returned to his seat but not without first shouting "Keep my wife's name out your f**ing mouth!" If you don't recall the situation I'm describing, it happened in 2022 during the 94th Academy Awards. Although Smith apologized, the consequences to him were quite drastic, with several of his projects cancelled or delayed. He also retired from the Academy and was banned from attending the Oscars for ten years.

This was almost a textbook example of **system 1** in action. The system 1 and system 2 framework, proposed by Daniel Kahneman and his collaborators, models how our brains process information. The framework is described in detail in Kahneman's book, *Thinking Fast and Slow*. System 1 is our fast, automatic thinking that is always on. It is unconscious, intuitive, and effortless.

And it is fast. It operates in milliseconds. System 2, on the other hand, is conscious, analytical, and effortful. It is also slow, requiring focus. System 1, which is constantly working, will generate impressions and intuitions. System 2 can either accept them, or can engage in a deeper process of questioning and analysis, which will sometimes override our initial intuition. While simple arithmetic problems like *2 + 2* are answered by system 1, we need system 2 to solve more complex problems, like *17 × 28*. When we rely solely on system 1, we may provide intuitive but incorrect answers.

One famous example of system 1 getting us in trouble is this simple problem proposed by Shane Frederick.[9] "A bat and a ball cost $1.10 in total. The bat costs $1.00 more than the ball.

How much does the ball cost?" If we let system 1 do its thing, we'll quickly answer ten cents. System 2 would easily be able to compute that if the total was $1.10, the bat would cost $1.05 and the ball just five cents. But system 2 is lazy, and we need to intentionally activate it. Sometimes, system 1 engages as a protection mechanism when we feel we are under attack. This was likely what happened in the Will Smith situation.

Chris Rock joked about Jada being bald. She suffers from alopecia areata, an autoimmune disease that causes

---

9. Cognitive Reflection and Decision Making, S. Frederick, The Journal of Economic Perspectives, Vol. 19, No. 4 , pp. 25-42, Autumn, 2005.

hair loss, and she had previously spoken publicly about the emotional toll of the condition. System 1 comes from our hunter-gatherer days, and is quick to engage. While it worked really well when we needed to run away from predators, it may put us out of our jobs in modern day social interactions. We live in a complex world and need system 2. When we think about the inner workings of GenAI, the transformer blocks with self-attention and feedforward phases behave much more like system 1. The parameters we encode in the model are the patterns we learned from the training data. Mostly, these models can repeat what they learned.

Researchers often call LLMs stochastic parrots.[10] Stochastic because the tokens are selected based on probabilities, and parrots because they claim these models don't really "learn" anything. Instead, their parameters encode the knowledge from all the documents in the training set. They know a lot because they read for 30 thousand years straight and memorized all concepts, storing them as model parameters in matrices. However, they lack a system 2, and cannot generalize the concepts and produce something truly novel. The fact that GPT-4 Mini could solve the math problem I posed to it when the numbers were 1, 2, 3, and 4, but could not answer the same problem when the numbers were 17, 18, 19, and 20 corroborates this idea. However, we do have evidence that

---

10. https://en.wikipedia.org/wiki/Stochastic_parrot (Last accessed in April 2025).

some of the answers these large models generate seem to be novel, or at least novel combinations of old ideas.

As we know that harder problems require deeper "thinking," some new models are now being extended with chain-of-thought (CoT) reasoning. As I write this, OpenAI models O1 and O3 are already trained with CoT. The idea is quite simple. Instead of training the model to produce an answer, we force the model into system 2 thinking by asking it to produce an explanation before producing the answer. This is a trick that forces the model to break down the problem into smaller steps, and use the intermediate outputs of these steps as new inputs. Let's consider a simple example of how CoT works. With no CoT, if we ask a problem like, "what is the average population of an American state?," a GenAI system may just present us with a plain answer, like "the average population of a U.S. state is 6.6 million people" without providing an explanation. With CoT, the model will first try to produce an explanation. The response would look something like the following: "the U.S. has fifty states. The country's population is estimated at 331 million people. To calculate the average population by state, we need to divide 331 million people by 50 states, giving us 6.6 million people per state, on average."

Forcing the system to add an explanation is similar in nature to activating our system 2 mode when we face a harder problem. Let's again revisit our favorite problem "if *1 + 1 = 2, 2 + 2 = 8, 3 + 3 = 18*, how much *4 + 4* would be?" We need to intentionally engage our system 2s if we want to solve this. The same is likely true for GenAI systems, although the bigger models may have seen this problem before and memorized it in the model parameters. Please remember models do not have calculators. All they know how to do is predict the next token. With CoT, their goal is first to explain the problem. A plausible explanation, as we've seen before, is:

$$(1 + 1) \times 1 = 1$$
$$(2 + 2) \times 2 = 8$$
$$(3 + 3) \times 3 = 18$$

This explanation becomes an intermediate solution that is then used to produce the output. By looking at these tokens, it is not far-fetched that a GenAI system could produce *"(4 + 4) × 4 ="* as next tokens. It may not know how to solve that either, but it again could produce intermediate results to explain the solution.

$$(4 + 4) \times 4 = 4 \times 4 + 4 \times 4$$

And it is very likely that it has learned during training that *4 × 4 = 16* and *16 + 16 = 32*. By forcing an explanation, it breaks down the problem into smaller subproblems. And the tokens produced as output for each of these subproblems are used as intermediate inputs. There is a lot of information encoded in the model parameters, and CoT is a way to access some of these connections. Without CoT, the only way to generate answers is using natural language in the prompts. CoT forces the system to produce explanations, which in turn generate intermediate results. It is like extracting more information for each problem, similarly to what our brains do when we engage system 2.

All the tokens, intermediate and final, are still produced by the transformer blocks we discussed before. The only difference is that with CoT, we trained the model to work harder. What we have learned by using GenAI models so far is that there are tasks they perform well using the straightforward next token algorithm. Grammar is one of these tasks. Old school AI models relied on rules, but we've realized that language is too complex and these rules have lots of exceptions. It is better not to try to encode rules and their exceptions and rely solely on next word prediction. After all, these models have read for thirty thousand years straight, and learned a lot or all of these exceptions. This is the same for humans. Grammar knowledge can, for the most part, be handled by our intuitive system 1. Most of us can speak intuitively, without the strain of engaging system 2.

Solving math problems and coding, on the other hand, cannot be easily done with system 1 alone. Even the smartest person on earth would have to engage system 2 to solve a math problem or come up with a novel algorithm. It is very likely our system 1 can take care of facts we have seen over and over, such as *4 + 4* or *5 × 7*. The larger language models are at an advantage over humans on that task, since they have seen more data and memorized much more than any of us. For anything novel, we need system 2. CoT is the GenAI version of that, at least for now. It is possible that we may find other ways to extract more knowledge out of model parameters when posed with complex tasks. Our hope is that by tuning the systems better for complex tasks, such as math and coding, we'll improve their analytical skills and their performance for a wide range of other tasks. It's like believing that someone that knows how to solve complex physics problems would, when provided with all the relevant data, be also able to reason about economic and geopolitical issues.

One other important characteristic of GenAI systems that we didn't spend much time on is their ability to learn from mistakes. This comes from the reinforcement learning from human feedback (RLHF) piece. It is like a human with a great growth mindset. And the good news is that computers don't feel sad when you tell them they did something wrong. They will gladly take your feedback and adjust the parameters to make the model better. It is important to realize that although interfaces such as ChatGPT will ask for our feedback, it won't be used to update the current version of the model. Remember that P in GPT stands for pre-trained. The parameters are already computed. Any feedback we provide when using GenAI tools will be used to update a new version of the model, the next time it is trained.

During training, the model is first computed without user feedback, using the entire training data. After that, users issue prompts and rate the responses, producing a much smaller training set that is used to update the model parameters. Based on these ratings, we can also build a prediction model that will evaluate the quality of answers. Given that, we can run a much wider set of prompts and let the system tune itself without the need for user input, further updating the model. This is similar, in spirit, to how AlphaGo learns how to play games just by playing against itself and trying to increase its best score.

Before we conclude the technical deep dive on GenAI, there is one last topic we need to cover. Some of the new systems support the integration with external tools, which provide added functionality on top of the generative models. ChatGPT, for instance, has integration with WolframAlpha, an external system specialized in advanced math computation, and Expedia for travel bookings. These tools can also be viewed as optimized system 2s for specialized tasks. The good news is that we can build them to efficiently solve the task at hand. The downside is that, while the GenAI model is generic, these external tools are not. They need to be custom built. AI purists think it may be possible to improve the generic GenAI models, by using more training data and storing more parameters, or by using novel techniques, such as CoT, so that we don't need some of these external tools. **Artificial general intelligence**, or **AGI**, would be a system that is generic, not requiring external specialized tools, and can do everything humans can, faster and more accurately. There is currently no clear consensus of what AGI really means, with several books dedicated to the subject.

Lately, the popular term **agent**, also known as **digital agent** or **AI agent,** simply means a software program that uses AI models to perform automated tasks, such as daily scanning several news sites, summarizing their content using GPT-3, and sending you an email with the summary. Many believe digital agents will eventually perform a lot of our daily tasks. In his recent book, *Superagency*,[11] LinkedIn cofounder Reid Hoffman and his coauthor, Greg Beato, argue that AI and digital agents will improve human agency to a new level of "superagency," allowing us to make better decisions in a more efficient way. They also claim that the way OpenAI releases new versions of their systems, through **interactive deployment**, allows them to minimize the negative impact of the technology by gathering early feedback. Interactive deployment will enhance our chance to build AI systems that augment our humanity and that will eventually lead to superagency, and should be the way we evolve all AI systems.

Now that we know a lot more about how GenAI works under the covers, let's step back and revisit how knowledge workers are using these systems to improve productivity.

---

11. *Superagency: Empowering Humanity in the Age of Ai,* Reid Hoffman and Greg Beato, Authors Equity, 2025.

The physical system "human beings"

By now, we have a good understanding of what GenAI can and cannot do. It should be clear that we don't have any AI system currently that is smarter than humans or that can dominate earth. The best AI systems we currently have only know how to predict the next tokens well. GenAI systems are not even capable of discerning good references from fake ones. And this is by design. There is nothing in their architecture that tries to identify reliable sources, like PageRank does. While doing the research for this book, several times I've tried to use GenAI tools and got links to references that don't even exist. In the end, I reverted back to good old web search in many cases. We are still a few innovations away from AGI, if we ever get there. However, even if we don't ever get to full AGI, the GenAI tools that we have today already can do a lot. They can really boost productivity in several domains. Let's go back to our personal usage effectiveness formula:

PUE = Total work / Total value-producing work

GenAI improves overall value-producing work (VPW) for many fields. Most of us can become more efficient by using GenAI tools to automate the labor intensive part of our jobs. When I was in college, studying computer science in the early nineties, one of my professors used to say "you can solve the problem up to the point a computer would be able to take over and complete the calculations." It resonated with me back then. And now it is more and more a reality that we can focus on the creative parts of our jobs, while computers can automate the tedious work. In *Bullshit Jobs,* late anthropologist David Graeber uses the term **duct tapers** for jobs that can be fixed permanently. For instance, instead of taking care of the leak in your bathroom, you instead hire someone to manage the bucket and empty it when needed. These are classical examples of jobs that can be automated away through technology.

We are also raising the bar on the quality of the work we expect to see. This has been happening for a few decades, since the personal computer revolution, and for many fields. But now the AI systems are driving quality up faster than ever before. The first spell checkers in the eighties made articles with spelling mistakes less acceptable. Then came automatic grammar corrections. Now, GenAI can provide restructuring suggestions that enhance comprehension, or even write the full text for you once you provide the talking points. Even though I mainly work with engineers who may not be super talented writers, I now expect to read technical reports that are concise, clear, and of course, have no grammar or spelling mistakes. It is also likely that, with the current tools, it takes less time for them to write these better technical reports than previous generations did in the past. And since technical reports and presentations are not the main part of their jobs, engineers now have more free time to do what they love the most – engineering.

Viewed this way, it is a win-win situation for both humans and computers. Engineers can focus on engineering, journalists can focus on investigations, lawyers on litigation, while computers do the matrix multiplications and number crunching they love doing. Or at least I hope they love, as it is the only thing they know how to do. We raise the bar on quality and increase productivity at the same time. And everyone is happy. We'll produce more value more efficiently, increasing VPW and consequently PUE globally. Such gains can then be reinvested, to further innovation efforts. As shown in Figure 12, the virtuous cycle for organizations requires investing in technologies that produce efficiencies and combat deterioration, such as spell checkers and GenAI, and use the gains to invest in technologies that will combat disruption, such as new payment methods and new online shopping tools.

However, as labor historian Jason Resnikoff described in his insightful book, *Labor's End*,[12] if history is any indication, technology and automation many times "degrade" jobs, transforming them into simpler tasks. For example, the job of a skilled warehouse worker may be broken down into mechanical steps, which may eventually be completely automated. The unchecked

---

12. *Labor's End: How the Promise of Automation Degraded Work*, Jason Resnikoff, University of Illinois Press, 2022.

use of technology on the labor force ultimately led to the abolition of manual skilled work from industry. Acemoglu and Johnson in *Power and Progress* also emphasise a related point. Technology and automation may negatively impact employee experience without significant gains in efficiency. For instance, scheduling software may optimize the number of workers in a store based on predicted customer demand, causing instability in the employee schedules in the name of efficiency. Surveillance technology is another example that negatively impacts the employee experience. As we are still in the early days of using AI pervasively in the workplace, we should act now to make sure history doesn't repeat itself.

I understand that chief financial officers (CFOs) and business leaders all over the world are looking into AI as a way to reduce costs, betting on the productivity gains it produces for knowledge workers. My own experience with **vibe coding**, the new term used for building software with the help of GenAI tools, is that it greatly helps in many scenarios, but it requires human assistance to correct mistakes and intervene when AI is not able to progress on its own. As reported in a New York Times article from June 2025,[13] GenAI helps experienced developers more than inexperienced ones. In that article, developers also claim that their work is becoming more

---

13. https://www.nytimes.com/2025/05/25/business/amazon-ai-coders.html (Last accessed in June 2025).

and more like assembly line work. I don't personally subscribe to that opinion, and I do like vibe coding. However, I do understand the code that GenAI is producing and I know how to guide it when it gets stuck. Inexperienced developers may not be able to do so in some scenarios. And they will never become experienced if they don't learn on the job.

An important point about software development is that programming languages allow us to write code that is precise and concise. Figure 2 is a very accurate description of bubble sort in just 15 lines. Vibe coding is the use of natural language, like English or Portuguese, to describe algorithms in prompts instead of describing them using programming languages. It works really well for a lot of scenarios, but for complex problems, nothing beats the precision of a structured programming language. As we discussed in Chapter 2, English descriptions can be vague and "do not study" is very different from "do not study here." Growing up, my favorite computer science book was Hopcroft and Ullman's book on formal languages,[14] as I value the precisiness in communication, especially when talking to computers, lawyers, and restaurant waiters. It would take a lot of writing to describe all the details of how to compute PageRank fully in English without any formulas or the formalism of programming languages.

---

14. *Introduction to Automata Theory, Languages and Computation*, J. Hopcroft and J. Ullman, Addison Wesley, 1979.

Kenton Varda, who is a very experienced software developer, being one the original developers behind Google's widely used Protocol Buffers library and other successful projects, recently published his efforts to create a new complex software using vibe coding.[15] This is a very good case study because Varda stored the history of all the 50 or so prompts he created during the two months he worked on the project. He started as a skeptic, but was convinced otherwise: "I was trying to validate my skepticism. I ended up proving myself wrong." However, there are key points to highlight: (1) Varda, being a very experienced developer, was able to unstuck the AI whenever it was not making progress; (2) his prompts contain very detailed descriptions, including code fragments that precisely described what he wanted; and finally (3) he had to do a lot of manual intervention. In the history log, 9 out of his last 11 changes start with the word "Manually," indicating he was completely bypassing AI.[16]

Given these reasons, I feel that CFOs shouldn't be rushing to cut jobs based on the promise of productivity gains from vibe coding. Additionally, in my 20+ years of working in the field, I was never part of a company or a project in which we had more developers than work to do. For most companies, it will be a while until the

---

15. https://www.maxemitchell.com/writings/i-read-all-of-cloud-flares-claude-generated-commits/ (Last accessed in July 2025).
16. https://github.com/cloudflare/workers-oauth-provider/commits/main/?before=a6e3e06c2642e0fd4c185374753201cff-c21ce8a+105 (Last accessed in July 2025).

productivity gains are so great that we can flip that imbalance. And even if we do, we can always apply the gains in productivity towards innovation, which in most cases generates more value than what can be offset by cost cutting initiatives.

In April 2025, a leaked internal manifesto from Shopify CEO, Tobias Lütke, went viral.[17] In his memo, he mentioned that employees should now justify why AI cannot be used to solve the problem before asking for new resources, be it more time or more people. This has been wrongly interpreted by many media outlets as AI replacing humans, which I don't believe was Lütke's original intention. He was simply mandating that people in his organization learn the AI tools and use them effectively to, in his words, Get S** Done (GSD). This is his way of saying, let's increase our percentage of VPW and, consequently, our PUE, by leveraging AI. As long as the company has room to innovate, humans will be needed. At least for the foreseeable future, AI doesn't know which novel algorithms to create, which bridges to build, which stories to report on, and which cases to litigate. Humans do. And humans also develop the AI tools that make computer scientists, civil engineers, journalists, and lawyers more efficient.

---

17. https://www.forbes.com/sites/douglaslaney/2025/04/09/selling-ai-strategy-to-employees-shopify-ceos-manifesto/ (Last accessed in April 2025).

Despite AI making all of us more productive, the number of problems we still need to solve seem to be increasing as well. We may be living in an era of abundance of productive resources, given the investment in education and technology we made in the previous decades and continue to make. However, we still don't have line of sight in containing the world's biggest problems, including global health, climate change, and inequality. Some of these problems are increasing. To make real progress on them, we need not only to continue our current investments in education and technology, but we may need to accelerate them even further. AI is a key enabler and may be how we cross the line in some of these areas. The 2024 Nobel Prize in chemistry, for instance, was awarded to Prof. David Baker from the University of Washington and Google researchers Demis Hassabis and John Jumper, for their work on proteins. Their work heavily relied on computer systems and AI tools. From the press release[18] "Life could not exist without proteins. That we can now predict protein structures and design our own proteins confers the greatest benefit to humankind." Another example is agriculture.[19] AI can connect farmers globally with timely insights, such as early disease warnings, democratizing expertise and enabling more predictive, data-driven decision-making. It includes AI

18. https://www.nobelprize.org/prizes/chemistry/2024/press-release/ (Last accessed in April 2025).
19. https://www.weforum.org/stories/2025/06/ai-food-systems-agricultural-revolution/ (Last accessed in August 2025).

guiding farmers when to irrigate, fertilize, or deploy beneficial insects to reduce waste, save labor, and help meet sustainability standards.

Like any technology devoted to increasing productivity, GenAI is already impacting the workplace. Mechanical tasks can now be automated. This allows workers to focus on more complex tasks. However, if they don't have the knowledge or experience to work on these more complex tasks, they are at risk of being automated away. This is likely true for every worker who doesn't have enough skills and experience, and performs only "system 1" tasks. GenAI is raising the bar for humans. AI helped chess players become more proficient over the years, being able to play against great AI opponents whenever they wanted. In the same way, we need to rethink what to teach new generations and how to accelerate their learning, so they don't become irrelevant. We have many relevant problems to solve, and we need capable humans that can positively contribute to society. I do agree with Reid Hoffman and Greg Beato that AI has the potential to increase our human agency. It can also raise the bar on the quality of the work we expect and increase our productivity. But we have to do it deliberately, preparing the new generations to be productive and engaged.

AI is all around us. We are already living the AI revolution. A deep understanding of how GenAI works provides us with valuable insights and guides us in the discussions we need to have. The next revolution is quantum, as we'll see in the next chapter.

7.

Quantum computing

*Roger Rabbit, spooky action, qubits, bogosort
revisited, and our future*

"Animation follows the laws of physics—unless it is
funnier otherwise."

Art Babbitt

A body suspended in space will remain suspended until
becoming aware of the situation. This means that a character that runs off a cliff will keep running as if nothing
happened, until looking down and realizing she is mid-
air. At that point, she will adhere to the laws of physics
and start falling down. Some bodies can go through a
solid wall, others cannot. A character who paints what
resembles the entrance of a tunnel in a flat wall, to trick
his opponent, is generally in trouble. His opponent will
likely get through the solid wall, while he will smash
into it. Sometimes the shock will flatten his body and
he will have to think about some inventive way to inflate
himself back to normal. Cats, however, are lucky. They
can be sliced and slayed in many ways, but they can never
be destroyed.

According to late Walt Disney Studios animator Art
Babbitt, cartoon characters follow the laws of physics

most of the time. Deviations happen only when it makes the situation funnier.[1] Esquire originally published "The Cartoon Laws of Physics" in 1980, but these laws have been modified and republished several times.[2] Since we have reviewed how to classify cat videos in the last chapter, it is super important for us to pay close attention to this corollary from law number eight from that article: "VIII: Any violent rearrangement of feline matter is impermanent. Corollary: A cat will assume the shape of its container." Even with modern AI techniques, it is very hard to identify cats inside bottles.

Cartoon physics is as old as animation itself. In 1956, Walt Disney spoke about the "plausible impossible" in an episode of the Disneyland television program. It has even been explicitly described in movies. Roger Rabbit himself is unable to escape handcuffs for most of a scene in the 1988 American fantasy comedy film *Who Framed Roger Rabbit*. However, when Eddie Valiant was trying to cut the handcuffs, Roger was able to free himself to hold the table still. Surprised, Eddie asked, "Do you mean to tell me you could've taken your hand out of that cuff at any time?" "Not at any time!," replied Roger, "Only when it was funny!"

---

1. https://en.wikipedia.org/wiki/Cartoon_physics (Last accessed in May 2025).
2. The Cartoon Laws of Physics, IEEE Institute, vol 10, 1994, available at https://web.archive.org/web/20120314215930/http://remarque.org/~doug/cartoon-physics.html (Last accessed in May 2025).

Outside the world of animation, the laws of physics are very real. And the only reason cartoon physics is funny is because we understand gravity and the laws of mechanics very well. If we didn't, we would be oblivious to the fact that they were being violated. We have a very good intuition about the physical world and the laws that govern it. The same is not true when talking about the world of the very small – the subatomic world of quantum physics. In this chapter, I'll describe quantum computing, but before we go there, we need to understand a few interesting properties of quantum physics. Unlike the world of cartoons and our physical world, it is much harder for us to understand what we cannot see and have direct interactions with. This was true even for Einstein and the physicists who worked on the early days of quantum physics.

Einstein was spooked

I mentioned before that one of the differences between virtual machines (VMs) and physical servers, is that you cannot kick a VM when you get frustrated. Before cloud computing, the server was conveniently placed under your desk, ready to be kicked when your program didn't produce accurate results. The VM, on the other hand, is running in a remote data center. You cannot kick it because the aptly named **principle of locality** states that all actions and their side effects are local. Kicking the server won't fix your code for you. Believe me, I've tried. But It will emit a sound and may cause the server to tumble and damage its hard drive. These are local effects. It likely won't cause your officemate's computer, located a few feet away, to also tumble, and it certainly won't affect the server that is running VMs, in a data center 2000 miles away.

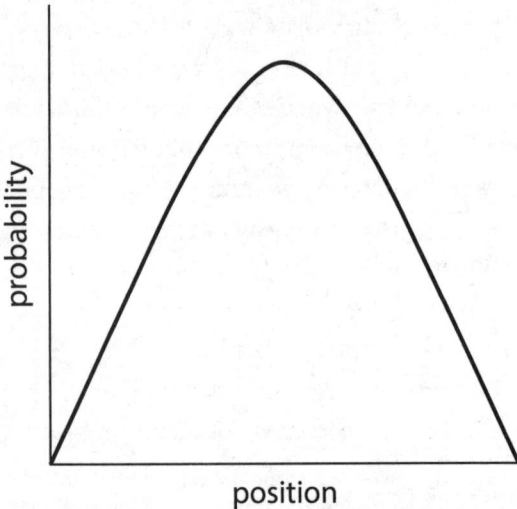

Figure 20. Sample wave function, which indicates the probability of the particle being at each position at a given time.

When we kick something, the force is locally applied. Even gravity is transmitted amongst objects through the principle of locality. It is transferred at a finite speed, as nothing travels faster than the speed of light.[3] There are

---

3. *When Einstein Walked with Gödel: Excursions to the Edge of Thought*, Jim Holt, Farrar, Straus and Giroux, 2018 (This is a fantastic book. Although I used different sources for this chapter, most of this section comes from what I learned from Jim Holt's book).

no instantaneous, non-local, actions in the world that we can directly observe and experience. In the world of atoms and subatomic particles, which is governed by quantum physics, the story is a bit more complicated. A particle moves according to a wave function that describes the probability of it being at any point at a given time. Think of an electron orbiting an atom, or an airplane going from New York to London. The wave function gives the probability of where in its trajectory the particle is, like there is a 10% chance that the plane is over Long Island right now. Figure 20 shows an illustration of a wave function, which assigns a probability for each position. As shown in the figure, It is a smooth function, without discontinuities or jumps, in the same way the plane doesn't skip from one continent to another without first flying over the Atlantic ocean.

In the subatomic world, we don't know where a particle is exactly at any given point. The only thing we know is that it adheres to its wave function. The term quantum physicists use to represent this is **superposition**, meaning that the particle could be in one of many positions. When we measure the particle position, however, the wave function collapses and the particle will be in exactly one location, out of the many predicted by its wave function. The simple act of measuring the particle's location causes the function to collapse. Perhaps the most well-known thought experiment that explains the concept of quantum superposition is Schrödinger's cat, devised by physicist

Erwin Schrödinger in 1935.[4] The cat, locked inside a box, is in a state of superposition, being either dead or alive depending on a subatomic event that may or may not occur. The occurrence of that event would poison the cat. Before we looked inside the box, we wouldn't know if the cat was poisoned or not, and he would be in a weird superposition between life and death, which is a paradox. This experiment came from a conversation between Schrödinger and Einstein to illustrate the problems they saw with the prevalent quantum mechanics interpretation.

Many mistakenly believe that Einstein had a problem with quantum mechanics's randomness, perhaps due to his famous quote "God does not play dice." However, it wasn't the randomness per se that bothered him. He was convinced that theory didn't provide a complete description of what was truly happening in the physical world. And violating the principle of locality was key to his argument. In quantum physics when two subatomic particles collide, they almost always become **entangled**.[5] Entanglement is the quantum mechanics term to describe that the particles are now linked together. One of the consequences of this entanglement is that measuring one will instantaneously

---

4. https://en.wikipedia.org/wiki/Schr%C3%B6dinger%27s_cat (Last accessed in May 2025).
5. *What Is Real?: The Unfinished Quest for the Meaning of Quantum Physics,* Adam Becker, Basic Books, 2019 (Also a fantastic book, especially if you want a deeper understanding of quantum physics and its history. My understanding for this section also came largely from this book).

fix the result we'd get when reading the other, no matter how far apart they are. We call this **quantum non-locality**. Einstein called it "spooky action at a distance."

Imagine you and a friend are each given a pair of magic gloves, one left and one right, sealed in identical boxes. The catch is, you don't know which glove is in which box. You take your box to Japan, and your friend travels with theirs to Brazil. When you finally open your box in Tokyo and see a left glove, you instantly know your friend in Rio has the right one. No matter how far you are. But in quantum entanglement, the situation is stranger: it's not that the gloves were left and right all along and you just didn't know. It's that they weren't decided until one of you opened the box. As soon as you look, reality "chooses," and that choice somehow travels faster than light to your friend's glove, instantly fixing its identity too. Even though you are in different countries. Spooky indeed. Einstein was a "realist." He believed science should provide an accurate and intelligible description of the world. In his view, quantum mechanics, or at least the **Copenhagen interpretation**, championed by Niels Bohr and his students, including Heisenberg, was an incomplete description of the world.

Another misconception about Einstein relates to his Nobel Prize. It was not awarded for the theory of relativity and the famous $E = mc^2$ equation. His 1921 award was for his pioneering work on quantum physics, specifically, his work on the photoelectric effect. So it is not true that Einstein was against quantum physics, a field he helped create. His concerns were the non-locality and the measurement problem, which caused the waves to immediately collapse. Bohr, contrary to Einstein, was an "instrumentalist." He didn't care about the spookiness of the wave collapse. He and his Copenhagen companions saw no problem with the split in which Newtonian physics describes our macro world while quantum physics describes the subatomic world. And in this view, there was no measurement problem, as there is no reason to worry about the position of particles when we are not looking. To Bohr, "there is no quantum world," as he famously put it.

Einstein, wanting a more complete and less voodoo-like interpretation, spent the last decades of his life debating with Bohr and challenging the Copenhagen interpretation. He proposed several experiments that highlighted his issues with non-locality and the wave function collapse. The simplest is known as Einstein's boxes. Let's say we confine a single electron to a box. Before any measurement, the electron will be happily moving inside the box according to its wave function. At that point, we put a partition in the middle of the box. If we do this carefully, the electron's wave function will split into two. One will be confined into the left of the partition and the other will be on its right. Since we still didn't do any measurement, the electron could be in either of these two parts with equal probability. We then separate the two halves of the box, sending one to Paris and the other one to Tokyo. Like in the example of the gloves above, when the physicist in Tokyo opens the box and detects the presence of the electron, the simple act of measurement would cause the wave function of the box in Paris to collapse at the same instant, like the two boxes were somehow connected. The Copenhagen interpretation saw no problem with this experiment. Their proponents dismissed not only Einstein's boxes but all his other attempts to highlight the inconsistencies of the theory.

The Copenhagen interpretation can be summarized as "Don't look: waves. Look: particles." And its underlying math has been very successful as a tool for several domains. It explains the entire discipline of chemistry, including the periodic table. It enabled the discoveries of the atomic bomb, silicon transistors, LEDs, and lasers, just to name a few. It is one of the most successful theories in science.[5] So it is really strange that while you and I are formed of atoms, and those are governed by quantum physics, it isn't able to describe us. Von Neumann accepted this duality. He wrote in his quantum physics textbook that "we must always divide the world into two parts, the one being the observed system, the other the observer. Quantum mechanics describes the events which occur in the observed portion of the world, so long as they do not interact with the observing portion." His view was aligned with the Copenhagen interpretation, and he made the observer the one responsible for the wave function collapse. At this point, I'm almost tempted to believe that the law that makes cartoon characters be subjected to gravity only when they realize that they run off a cliff may be related to the quantum measurement problem.

One of the reasons Einstein did not like non-locality is one of the most fundamental principles in special relativity theory. Nothing, no single object or message, can travel faster than the speed of light in vacuum. Violating this would have very strange consequences, like being able to go back in time. Luckily, although quantum non-locality allows instantaneous changes at a distance, it does not allow for time travel and does not violate special relativity. The intuition behind this limitation is simple. As the results of measurement are random, quantum non-locality cannot be used to transmit messages instantaneously. Still, to Einstein, despite all the tremendous success of quantum physics as a vehicle to advance science, the Copenhagen interpretation was incomplete. The instrumentalist Bohr was happy, the realist Einstein was not.

By now you must be very anxious with your unresolved duality, with waves governing your atoms while you, as a whole, behave according to classical Newtonian physics.[6] We'll go back to your problem, and to Einstein's dilemma later. I'm sorry to leave you in a cliffhanger. Just

---

6. New age esoterics since the 70s, on the contrary, seemed to have dodged this anxiety by embracing quantum physics to justify all sorts of mystical thinking. The universe supposedly responds to your inner state, manifesting corresponding external outcomes. They would say quantum physics has "proven" that by observing something, we change it. Therefore, by focusing your thoughts on a goal, such as a romantic interest, you collapse the quantum wave function into the reality you desire. Perhaps some of you will have been exposed to "quantum physics" through some of this new age philosophy. Books like *The Tao of Physics* (by Capra, 1975) and *The Dancing Wu Li Masters* (by Zukav, 1979), and movies like *The Secret*, may have contributed to much of this disinformation.

keep running as nothing happened, and please, don't look down. Let's switch gears and use some concepts we've learned so far to start diving into quantum computing. We'll begin with the definition of **qubits**, the equivalent of bits in the quantum world. We'll then move on to quantum algorithms.

Classical bits and qubits

So far, we have talked about computers as processing functions over strings of bits. For an arbitrary function $f(x)$, we could have $f(1010001) = 101$. These functions can be interpreted as integer functions, since we can easily map bit strings to integers. In this example, 1010001 and 101 represent 81 and 5, respectively, if we simply map binary numbers to their decimal representation. We could then interpret that the value we get by calling $f(x)$ with input 81 is 5, or more compactly $f(81) = 5$. Additionally, hopefully I also convinced you that any algorithm that is implementable in a computer is an integer function, such as $f(x)$ above. In this example, $f(x)$ could be something simple, such as $f(x) = (x - 1) / 16$, but it also could be bubble sort, list intersection, Huffman coding compression, matrix multiplications, transformer blocks or anything else. Anything we ask a computer to do, either through algorithms written in programming languages or GenAI prompts, ultimately maps to integer functions that operate over bit strings.

But what are bits? As we've seen before, bits stand for **b**inary dig**its**, the smallest unit of information in a digital system. We know each bit is either 0 or 1, similar to a light switch. This is our logical interpretation of what they mean. Physically, in modern CPU and GPU hardware, bits are stored and processed using transistors, which behave like tiny electronic switches that can be on or off. For instance, 0 can be represented by 0 volts while 1 is represented by supply voltage, like 1.2 volts. These voltage levels flow through **logic gates**, also built from transistors, which implement all the possible functions we apply to bits, and ultimately, everything we can do with computers. Let's consider a simple example, adding single bits. The expected values are:

```
0 + 0 = 00
0 + 1 = 01
1 + 0 = 01
1 + 1 = 10
```

We need two bits for the output as adding *1 + 1* is 2, which is represented in binary as 10. 1 is the carry over bit. To implement this simple addition we need two logic gates, one for the sum and one for the carry over. We can write these as two functions:

```
sum(A, B) = XOR(A, B)
carry(A, B) = AND(A, B)
```

Function *sum*, is implemented by the **exclusive or** gate, or **XOR**, which returns 1 if the input bits are different and 0 otherwise. For instance, *XOR(0, 1)* is 1 while *XOR(1, 1)* is 0. Similarly, function *carry* uses the **AND** gate, which returns true only if both input bits are 1. Putting all this together, we'd have:

0 + 0 = 00, as sum(0,0) = XOR(0, 0) = 0 and carry(0, 0) = AND(0, 0) = 0

0 + 1 = 01, as sum(0,1) = XOR(0, 1) = 1 and carry(0, 1) = AND(0, 1) = 0

1 + 0 = 01, as sum(1,0) = XOR(1, 0) = 1 and carry(1, 0) = AND(1, 0) = 0

1 + 1 = 10, as sum(1,1) = XOR(1, 1) = 0 and carry(1, 1) = AND(1, 1) = 1

It may seem surprising, but all operations we do in a computer are implemented using combinations of a few basic logic gates, such as AND, OR, NOT, and XOR. Logic gates are built in transitions on integrated circuits, or chips. These are CPUs and GPUs we have in our modern servers. Hardware has evolved since the ENIAC days by packing more transistors into chips. We talked about Moore's law in Chapter 1. Calling it a law is a misnomer, as it is nothing more than an empirical observation by Intel's cofounder, Gordon Moore, in 1965, which states that the number of transistors in integrated circuits doubles every two years. Its implications are more processing power at lower cost and lower power consumption. Although it held true for many decades, Moore's law is reaching its limits post-2010, with transistors getting close to their physical limits of 1–2nm (nanometers) and cost per transistor no longer dropping as sharply as before. As Adam Becker puts it in his recent book, *More Everything Forever*,[7] nothing can benefit from an exponential scale in perpetuity.

As we'll see, quantum computers are not an evolution of classical computers. They are not going to help us extend Moore's law, as the goal is not to produce even smaller transistors. In fact, they would not be smaller at all. Integrated circuits for classical computers scale to billions of bits in a chip. Although we still don't have large scale implementations of quantum computers, they will likely have a much smaller number of qubits per chip. Currently, Microsoft seems to be leading the race with its Majorana 1 quantum chip. All going well, building the most scalable quantum processor with one million qubits is still a few years out. Logical operations in quantum computers are also slower and have a higher error rate, compared to classical computers. As physical qubits will always remain noise in every implementation of quantum computers, they need to be corrected to produce logical qubits. Table 2 today shows a comparison of bits and logical qubits.

7. *More Everything Forever: AI Overlords, Space Empires, and Silicon Valley's Crusade to Control the Fate of Humanity*, Adam Becker, Basic Books, 2025.

Criteria	Classical digital computer	Quantum computer
Information carrier	bit	qubit
Speed	picoseconds, or 0.000001 microseconds	10s to 1000s of microseconds
Error rate	10^{-15}	10^{-6} now, aiming for 10^{-18} in a few years
Scale	billions of bits per chip	10s of qubits now, aiming for thousands in a few years

Table 2. Comparison of classical digital and quantum computers. Take this with a grain of salt, since we still don't have a scalable quantum computer and these numbers may change drastically as we make progress in their development.

Although we'll have, at least initially, a lot less qubits than bits to work with, it may be plenty enough. As I mentioned, quantum computers are very different from classical digital computers. They are not going to be used to solve the same types of problems. But before we investigate which problems they will help us with, let's understand what qubits are. We've seen that bits are implemented by different voltages in transistors. There are few alternative ways qubits can be implemented, including polarization encoding of photons, the atomic level of trapped ions, and the spin of electrons. The properties of qubits and how they are used in the design of quantum algorithms will be the same, independent of the chosen physical implementation. Let's consider the spin of electrons as an example, shown in Figure 21, just to make the discussion more concrete. For the rest of the chapter we can largely ignore how qubits are physically implemented.

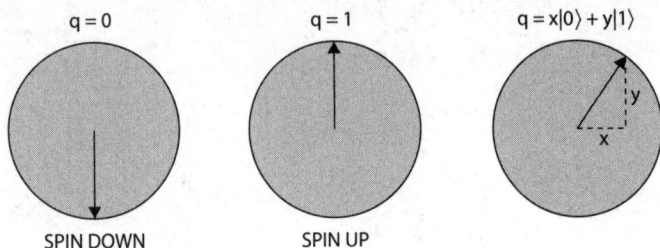

Figure 21. Qubits implementation using electron spins. Spin down means qubit q is 0. Spin up means it is 1. Unlike classical digital computers, qubits can be in a superposition state. In that case, reading the qubit would return 0 with probability x^2 and 1 with probability y^2.

All the quantum properties we have been discussing in this chapter apply to qubits. Before we read the value of a qubit, it will adhere to its wave function. Instead of being deterministically either 0 or 1, it will have different probabilities of returning these values. The simplest way to visualize a qubit's wave function is a circle with a radius of size 1 around the origin. The qubit can then be represented by a two-dimensional vector *(x, y)*, although quantum physicists prefer to write it as an equation $|q\rangle = x|0\rangle + y|1\rangle$, where the "ket" symbol "$\rangle$" indicates that these are qubits instead of regular numbers. This equation indicates that the probability of the qubit being 0 is

x^2 and the probability of it being 1 is y^2. These probability values are derived from the triangle shown in Figure 21. The qubit has no assigned value until we measure it. At that point, its wave function will collapse, and we'll read either a 0 or a 1, with probabilities x^2 and y^2, respectively.

One common misconception about quantum computers is that they can magically try every possible solution for a problem at once, simply because qubits can represent more than just a 0 or a 1. It is true that qubits have more information than classical bits, the values of x and y we have just discussed. It is also true that with n qubits we can process 2^n operations in parallel. But in reality, qubits are more like spinning coins that can be in a mix of heads and tails until you look at them. This mix, called a superposition, lets quantum computers explore many possibilities at once. But only in a very specific, structured way. You can't just read out all those possibilities at the end. When you measure a qubit, it still gives you a single answer, 0 or 1. The real power of quantum computing doesn't come from storing more information, but from the clever quantum operations, like interference and entanglement, that steer the system toward useful answers faster than classical computers for certain kinds of problems.

In the same way classical computers manipulate bits using logical gates, quantum computers operate with quantum gates. While logical gates mainly operate by manipulating the bit values, like *NOT(1) = 0* and *XOR(1, 0) = 1*, quantum gates operate by transforming the qubit vectors. There are gates to rotate and reflect vectors. A

Hadamard gate, for instance, puts the vector in **uniform superposition** by assigning equal probabilities for 0 and 1. Applying Hadamard to qubit $|0\rangle$ would rotate it to:

$$H(|0\rangle) = \tfrac{1}{\sqrt{2}}|0\rangle + \tfrac{1}{\sqrt{2}}|1\rangle$$

Quantum gates are very different from classical gates. The problems that benefit the most from quantum algorithms, achieving the biggest speedups, are those that can be efficiently solved with the vector manipulations provided by quantum gates. Next we'll see an example of a quantum algorithm in action. Although I've tried to simplify the presentation of these concepts, the next two sections are still very involved and parts of it may be hard to follow for non-technical readers. They are for illustration purposes only, and skipping them won't sacrifice the overall point of the chapter.

A quantum version of bogosort

In Chapter 4, we learned that NP-hard problems, like the knapsack problem we studied to understand how cloud services allocate VMs into physical machines, have no efficient, polynomial time implementation. However, we always have an efficient algorithm to verify the solution of an NP-hard problem. It may take a long time for us to find the solution of a maze, but if a solution is given to us, we can efficiently verify if it is correct. Another good example is sudoku. A traditional sudoku board is 9×9. A puzzle with 40 initial squares completed would give us 9^{41}

possible combinations to be tested, as shown in Figure 22. This is a very large number. Hopefully not all these combinations are valid and sudoku players have strategies to solve the puzzle without having to go through nearly as many options. Although solving sudoku is hard, verifying a solution is simple and efficient. We can write a simple algorithm that will go through the board in a single pass and verify that all the lines, columns, and blocks have all the nine numbers.

5		9				4		
7		8	3		4	9		
6		1				7	3	
4	6	2	5					
3	8	5	7	2		6	4	9
1		7	4		8	2		
2			1					4
		3		4			8	7
	7			5	3			6

Figure 22. Sample sudoku board with 40 starting positions. We have 9^{41} combinations left.

If our procedure to solve sudoku was to test all combinations, one by one, using our verification algorithm, it would take us about 10^{10} years in a modern day computer.

This is considering that our program was efficient and could process billions of board verifications per second. Bogosort from Chapter 1 tried a similar strategy for sorting lists. Here is its algorithm again:

1. Shuffle the list arbitrarily.
2. Check if the list is sorted.
3. If not, repeat step 1.

As in the case of sudoku, verifying if the list is sorted is efficient, so step 2 in our algorithm is fast. We just need to step through the list and check if all the elements are in order. However, since a list of n elements has $n!$, or n factorial, possible permutations, going through every option would take a long time. Although this is not a good way to solve sorting or sudoku, it may be the only solution in a few situations. If we are trying to guess a password, we may have no better option than trying all the options until we find the one that works. With n options, it would take us $\frac{n}{2}$ tries on average to find the answer. If we are unlucky, the correct answer may be the last one we try, so in the worst case it would take us n calls to the verifier program. But if we had access to a quantum computer, we would be able to do it in less steps. Computer scientist Lov Grover proposed the **quantum search algorithm**, known as **Grover's algorithm**.[8]

---

8. https://en.wikipedia.org/wiki/Grover%27s_algorithm (Last accessed in May 2025).

Formally, we have a verifier function that returns true if the input is valid and false otherwise. In the case of sorting, it returns true if we call it with correctly the sorted permutation. As we've seen, we have $n!$ possible permutations. We can assign an integer for each. For instance, with 3 elements, we'd have $3! = 3 \times 2 \times 1 = 6$ permutations. Let's consider input list $\{41, 5, 57\}$:

Permutation 1 = $\{1, 2, 3\}$. verify($\{41, 5, 57\}$) = false
Permutation 2 = $\{1, 3, 2\}$. verify($\{41, 5, 57\}$) = false
Permutation 3 = $\{2, 1, 3\}$. verify($\{41, 5, 57\}$) = true
Permutation 4 = $\{2, 3, 1\}$. verify($\{41, 5, 57\}$) = false
Permutation 5 = $\{3, 1, 2\}$. verify($\{41, 5, 57\}$) = false
Permutation 6 = $\{3, 2, 1\}$. verify($\{41, 5, 57\}$) = false

The verifier function would return true for only one of the $n!$ combinations, as only one of these permutations is sorted. In this example, function *verify* returns true for the third permutation and false for all the other ones. Permutation 3 is the only one that describes the correct order for the input list: the smallest element is in the second position, followed by the elements in the first and third positions. In a classical digital computer, all we can do is call *verify* for every possible input and hope for the best. Let's see how Grover's algorithm can do better. If we have a computer with n qubits, we'd assign a permutation for each of the 2^n possible qubits combinations, as follows:

Permutation 1. Qubit = $|00000000...0000\rangle$

Permutation 2. Qubit = $|00000000...0001\rangle$

Permutation 3. Qubit = $|00000000...0010\rangle$

Permutation 4. Qubit = $|00000000...0011\rangle$

..

Permutation 2^n. Qubit = $|11111111...1111\rangle$

We've seen that a single qubit can be represented as a vector in a two-dimensional space. Similarly, with n qubits, we'd have a vector in a high-dimensional space. A hyper space of 2^n dimensions to be precise. In the case of a single qubit we would always read either 1 or 0 as output. In the case of n qubits we'll read one of the 2^n possible combinations. We can represent these possibilities in equation form as:

$|q\rangle = w_1|0000...0000\rangle + w_2|0000...0001\rangle + w_3|0000...0010\rangle + ... + w_2^n|1111...111\rangle$

Each of the weights represent the probability of us reading a given qubit combination. For a single qubit, x^2 and y^2 represented the probability of us reading either 0 or 1, respectively. If we want to be precise, they are the probability of reading qubit $|0\rangle$ or qubit $|1\rangle$. In this general case with n qubits, the probability of reading, for instance, the third qubit combination $|0000...0010\rangle$ is w_3^2. The sum of these probabilities, the squares of all 2^n weights, is always equal to 1, as one we'll always read exactly one of the 2^n qubit combinations.

Quantum computers are powerful because their quantum gates allow us to perform operations on the high-dimensional qubit vectors. Rotating a vector in a space with 2^n dimensions, means changing the values of the 2^n weights in a single step. This is much more than classical digital computers can do. However, as we cannot directly read the weights, we need to design quantum algorithms that will make it very probable that we will read the qubit combination that contains our result. To do this, Grover's algorithm increases the weight for the single qubit that represents the right answer and decreases the weights for all the other possible $2^n - 1$ qubits. Let's see how Groover's algorithm solves our bogosort example.

Permutation 1 = {1, 2, 3}. Qbit = $|000\rangle$. verify({41, 5, 57}) = false

Permutation 2 = {1, 3, 2}. Qbit = $|001\rangle$. verify({41, 5, 57}) = false

Permutation 3 = {2, 1, 3}. Qbit = $|010\rangle$. verify({41, 5, 57}) = true

Permutation 4 = {2, 3, 1}. Qbit = $|011\rangle$. verify({41, 5, 57}) = false

Permutation 5 = {3, 1, 2}. Qbit = $|100\rangle$. verify({41, 5, 57}) = false

Permutation 6 = {3, 2, 1}. Qbit = $|101\rangle$. verify({41, 5, 57}) = false

We can write the qubit equation as:

$$|q\rangle = w_1|000\rangle + w_2|001\rangle + w_3|0010\rangle + w_4|011\rangle + w_5|100\rangle + w_6|101\rangle$$

As permutation 3 is the correct answer, Grover algorithm will manipulate the qubit, using gates that rotate it, so that w_3 gets close to 1 while the other weights get close to 0. Even if we are successful in doing that, we still could read a different qubit, as we are talking about probabilities. If Grover's algorithm outputs a qubit representing the wrong permutation, we can always re-invoke it. As the probability of reading the wrong qubit will be low, with just a few calls to the algorithm we are guaranteed to get the right output. But how does the algorithm work? I'll provide the intuition here, but there is a great video[9] from 3Blue1Brown, the same channel that I referred to for the embeddings and GPT-3 explanation, which provides a fantastic visual description.

We start by using Hadamard gates to position the vector to a uniform superposition state, in which all the qubits combinations have the same probability:

$$|q\rangle = \tfrac{1}{\sqrt{6}}|000\rangle + \tfrac{1}{\sqrt{6}}|001\rangle + \tfrac{1}{\sqrt{6}}|010\rangle + \tfrac{1}{\sqrt{6}}|011\rangle + \tfrac{1}{\sqrt{6}}|100\rangle + \tfrac{1}{\sqrt{6}}|101\rangle$$

As it is hard to draw a high-dimensional space, let's collapse it to 2D. We can always project a higher dimensional space in a lower dimensional one, in the same way we can project 3D structures into 2D for easier visualisation, as in the case of architecture drawings. Here we'll

---

9. https://www.youtube.com/watch?v=RQWpF2Gb-gU&t =1680s (Great video by 3Blue1Brown that explains Grover's algorithm in detail using animation. Last accessed in May 2025).

combine all the dimensions that don't correspond to the valid permutation, into a single, combined dimension, as shown in Figure 23.

The algorithm works by reflecting the vector multiple times, making it closer and closer to the vector representing the qubit that corresponds to the right permutation. In this case, we want to move $|q\rangle$ closer and closer to $|010\rangle$. As we've already seen, in quantum computers we can manipulate all qubit $|q\rangle$'s weights in parallel using quantum gates. In Grover's algorithm case, we'll invoke the *verify* function for each dimension. For each qubit combination we'll flip its weight value by multiplying it by -1 if *verify* returns true and we'll leave the weights intact otherwise, if *verify* returns false. Geometrically, this means reflecting $|q\rangle$ over the axis that is the combination of the other dimensions, as shown in Figure 23. In our example, it would change the weight of the third component, the one with the right permutation, to $-\frac{1}{\sqrt{6}}$. The new value of $|q_{new}\rangle$ after the initial reflection would be:

$$|q_{new}\rangle = \tfrac{1}{\sqrt{6}}|000\rangle + \tfrac{1}{\sqrt{6}}|001\rangle - \tfrac{1}{\sqrt{6}}|010\rangle + \tfrac{1}{\sqrt{6}}|011\rangle + \tfrac{1}{\sqrt{6}}|100\rangle + \tfrac{1}{\sqrt{6}}|101\rangle$$

It turns out that if we then rotate this updated vector $|q_{new}\rangle$ over the uniform superposition vector, it will get closer to $|010\rangle$. If we compute the angles betweens these vectors, we can show that we just need square root of the number of permutations move $|q\rangle$ close enough to $|010\rangle$. In this example we'd need $\sqrt{6}$ steps to get the right answer, as in each step we are reflecting the vector by an angle Θ close to $\frac{1}{\sqrt{6}}$. When we are done, its equation will be approximately:

$$|q_{final}\rangle \cong 0|000\rangle + 0|001\rangle + 1|010\rangle + 0|011\rangle + 0|100\rangle + 0|101\rangle$$

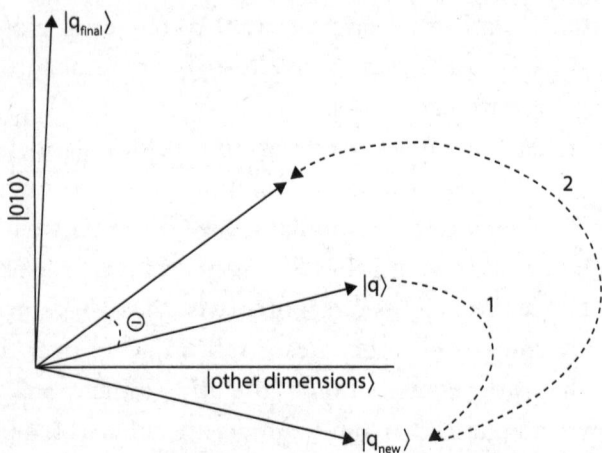

Figure 23. This figure shows the first step of Grover's algorithm. We start with the uniform superposition state $|q\rangle$. Each step of the algorithm involves two vector reflections. The first one reflects $|q\rangle$ over the hyperplane formed by the other dimensions, giving us $|q_{new}\rangle$. The second reflection moves $|q_{new}\rangle$ over the original uniform superposition state. These two reflections combined, move is an angle Θ closer to the final qubit $|010\rangle$. As Θ is approximately $\frac{1}{\sqrt{6}}$ in this example, after $\sqrt{6}$ steps will be aligned with $|q_{final}\rangle$.

This means that'd have a great chance of reading $|010\rangle$ as the algorithm's output, enabling us to correctly sort the list.[10] $\sqrt{6}$ is much better than the 6 steps that would be generally required in a classical digital computer. A research paper[11] in 1997 proved that it is impossible for a quantum computer to solve the quantum search algorithm in fewer steps than this, implying that Grover's algorithm is optimal. With 2^n possible answers, Grover's algorithm would require $O(2^{n/2})$ steps, using our big-O order notation for algorithmic efficiency. It is a substantial improvement, but not earth shattering. If we were trying to use Grover to discover an encryption key of 256 bits, we'd still need 2^{128} steps, which would take a very long time. So it is unrealistic to use Grover's algorithm for that task. This brings us to an interesting point. Since with n qubits we can manipulate a vector of 2^n dimensions in one step, why can't we solve quantum search in a single step?

The Grover's animation video I mentioned[12] provides a good discussion about this. The main reason is that weights are not available to us directly. If we could read the weight values we'd be able to solve quantum search in a single step. As that is not the case, we need to perform vector manipulations that will increase the probability of

---

10. Please see a visualization of Grover's algorithm at https://digitalagencybook.org/visualizations/grover-algorithm.

11. The strengths and weaknesses of quantum computation, C. Bennett, E. Bernstein, G. Brassard, and U. Vazirani, SIAM Journal on Computing. 26 (5): 1510–1523, 1997.

12. https://www.youtube.com/watch?v=RQWpF2Gb-gU&t=1680s (Great video by 3Blue1Brown that explains Grover's algorithm in detail using animation. Last accessed in May 2025).

us reading the right qubit as output. The reason Grover improves over the search in classical computers from $O(N)$ to $O(\sqrt{N})$ steps is that it allows us to geometrically get closer and closer to the correct vector using the "diagonals." If we think about a right triangle with base and height 1, the size of the hypotenuse would be $\sqrt{2}$. Classical computers can only search through the cathetus by, for instance, changing the bit sequence from 0011 to 0100 and invoking *verify*. The hypotenuse is not available in classical computers. Figure 24 highlights this difference between classical and quantum computers, which is the source of speedup we get from Grover's algorithm.

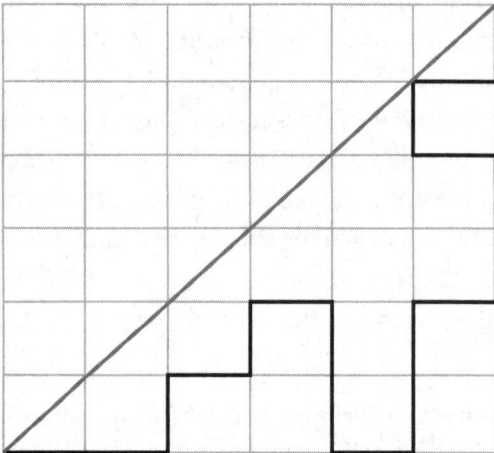

Figure 24. Illustration of a search algorithm in a regular digital computer, which only has access to the cathetus, and in a quantum computer, which can go through the diagonals. Being hand-wavy, this is where the speedup from $O(N)$ to $O(\sqrt{N})$ steps come from.

You may be asking yourself, why can't we just set the weight of the component for which *verify* returns true, like qubit $|010\rangle$ in our example, directly to 1 and all other weights to 0? This would allow us to solve any NP-hard problem in a single step. Unfortunately, we don't have a quantum gate that allows us to apply non-linear transformations directly. Setting the amplitude to 1 for the correct solution and 0 for others is **not unitary** in general, which is forbidden by quantum mechanics. A quantum operation must preserve the total probability amplitude (unitarity), and cannot be used to "zero out" wrong answers arbitrarily. Measurement, which in our case means reading the qubit, is non-unitary as it collapses the values and it is not reversible. In the absence of non-unitary gates, the best we can do, in the general case, is the $O(\sqrt{N})$ steps in Grover's algorithm. There are some problems, however, that allow us to have exponential speedups. You may have heard that quantum computers can break encryption efficiently. Let's examine how they do that.

Destroying internet security

In 1981, during his keynote speech at an MIT conference, Richard Feynman posed the question, "can physics be simulated by a universal computer? Can you do it with a new kind of computer – a quantum computer? I'm not sure… So I leave that open." In 1985, physicist David Deutsch proved that we could build a quantum computer that would be faster than classical digital computers for certain tasks. However, Deutsch didn't come up with any practical algorithms for the quantum computer. Nor did he build a computer. He just proved theoretically that it could be done.[5] But in 1994, American computer scientist Peter Shor proposed his famous Shor's algorithm. A quantum algorithm that can discover the factors of a large integer exponentially faster than in classical computers.[13]

When we hear about quantum computers being able to break internet security, we are most likely hearing about Shor's algorithm. To understand what it does, and how, we need to cover a bit of math. RSA, for Rivest-Shamir-Adleman, is an encryption system that operates based on large integer keys. In RSA, after we encrypt a message with a public key, which can be shared publicly, we can only decode it if we have the private key, which is a prime factor of the public key. The larger the number, the more computational power is needed to find the prime factors, and the more secure the encryption is. Although multiplication of two numbers is extremely efficient in classical digital computers, finding the factors, which means finding A and B for which $N = A \times B$, is slow.

Let's say you have a big padlock, and it's secured with a code that's based on multiplying two secret numbers together, like:

$$15 = 3 \times 5$$

---

13. https://en.wikipedia.org/wiki/Shor%27s_algorithm (Last accessed in May 2025).

This seems easy to undo. A computer could easily try all the number combinations until 15. But imagine you had a much bigger number like, 589,261. If you don't know the secret numbers, it could take a long time to figure it out, as we have an exponential number of combinations to try. RSA keys are much larger numbers, generally 512 bits or higher. Using a powerful modern server, it would take thousands of years of trying possible combinations to break an RSA key.

We could, of course, use Grover's algorithm, as we have a simple enough verification function. Just multiply the numbers. However, as we've already seen, Grover would not help much in this case. For 512 bit keys, it would only reduce the computational time from 2^{512} to 2^{256} steps, still taking way too long. Shor's algorithm does much better, providing an exponential speedup. It is the most prominent example of how quantum computers can achieve such larger speedups in certain scenarios. It is a clever quantum shortcut for factoring those big numbers. Like cracking the combination without trying every single one. It does this by turning a hard math problem, finding the factors, into a problem that can be more easily solved in a quantum computer, finding a repeating pattern.

Imagine being blindfolded in front of a piano. Someone plays a repeating combination of keys. Your goal is to figure out exactly how many keys there are in the pattern. Now, this might seem hard blindfolded. But what if you had super ears and you could hear all the notes at once, in the whole scale, and detect the pattern instantly? That's what the quantum computer does. Let's try to understand the details.

Again, the algorithm's goal is to find the factors of a larger number N. It relies on the fact that numbers that don't share factors, let's say A and B, can be written as:

$$A^P = m \times B + 1$$

for some power P and some multiple m. For instance, let's say $A = 7$ and $B = 15$. We'd have:

$7^4 = 160 \times 15 + 1$ (P = 4 and m = 160. I computed these values by hand since 7 and 15 are small numbers. Doing so for large numbers would require trying a very large number of combinations.)

This works for any numbers that don't share factors. Shor's algorithm exploits this property. It starts with a random guess, let's say G. Let's assume the public key N and the guess G are the inputs to the problem. Using the the trick we just learned, we know that:

$$G^P = m \times N + 1$$

This means that if we find P we can find the factors and break the encryption. I'm glossing over a few details for clarity's sake, such as the fact that P has to be an even number. The key insight Shor had is that P is the period of the function. For instance, with the $G = 7$ and $N = 15$ in example above, we'd have:

Reminder of 7^1 divided by 15 = 7
Reminder of 7^2 divided by 15 = 4
Reminder of 7^3 divided by 15 = 13
Reminder of 7^4 divided by 15 = 1
Reminder of 7^5 divided by 15 = 7
Reminder of 7^6 divided by 15 = 4
Reminder of 7^7 divided by 15 = 13
Reminder of 7^8 divided by 15 = 1
Reminder of 7^9 divided by 15 = 7
…

As expected, since we had already computed that $P = 4$, the period of the function is 4, as the sequence 7, 4, 13, 1, will keep repeating itself. These are like the repeating piano keys. If we can compute the period, by detecting the length of the sequence, we are pretty much done. However, computing this in a classical computer is extremely inefficient. We are better off trying to find the factors directly. Shor was able to devise a quantum algorithm that efficiently finds the period. For a public key of n bits, Shor's algorithm uses two qubits sequences of size n. And it uses a technique we haven't seen before. It entangles the two qubits sequences. For $N = 15$, and $G = 7$ as above, the expression for the entangled bits would be:

$$\frac{1}{\sqrt{256}} |0\rangle |7^1 \text{ divided by } 15\rangle +$$
$$\frac{1}{\sqrt{256}} |1\rangle |7^2 \text{ divided by } 15\rangle +$$
$$\frac{1}{\sqrt{256}} |2\rangle |7^3 \text{ divided by } 15\rangle +$$
$$\frac{1}{\sqrt{256}} |3\rangle |7^4 \text{ divided by } 15\rangle +$$
...

The second qubit values follow the periodic sequence we computed above, 7, 4, 13, and 1. If we read the second qubit, we'll get one of these four values, randomly selected as all the probabilities are the same. Let's say we get 4. It doesn't matter for the algorithm, as any of the four values will work the same way. As the two qubits are entangled, when we read the second qubit, the first one will collapse, and we'll only keep the values for which the second qubit is 4. Our expression would be simplified to:

$$|q\rangle = \frac{1}{\sqrt{64}} |1\rangle + \frac{1}{\sqrt{64}} |5\rangle + \frac{1}{\sqrt{64}} |9\rangle + \frac{1}{\sqrt{64}} |13\rangle + \ldots + \frac{1}{\sqrt{64}} |253\rangle$$

We kept only the components that lead to a reminder of 4. Please note that since we reduced the number of components by 4, we also had to adjust the weights from $\frac{1}{\sqrt{256}}$ to $\frac{1}{\sqrt{64}}$. We'd used $\frac{1}{\sqrt{256}}$ originally, as with keys of 8 bits we have 256 possible values. To find the value of P, Shor's algorithm relies on a quantum version of the Fourier transform, called the quantum **Fourier transform**, or **QFT**. Given a pattern of qubits, like $|1\rangle$, $|5\rangle$, $|9\rangle$, $|13\rangle$, etc. in our case, QFT uses unitary gates to give us a sharp peak at positions that are multiples of Q over P, where Q is the total number of possible values, 256 in our case, and P is the number we are looking for, which is 4 in our case. This means that QFT would boost the weights of qubits 0, 64, 128, and 192, and make the weights of every other component close to 0, leaving us with expression:

$$QFT(|q\rangle) \cong \tfrac{1}{\sqrt{4}}|0\rangle + \tfrac{1}{\sqrt{4}}|64\rangle + \tfrac{1}{\sqrt{4}}|128\rangle + \tfrac{1}{\sqrt{4}}|192\rangle$$

When we read the output, we'd read one of these four values, 0, 64, 128, 192 at random. Reading 0 wouldn't be very helpful, in which case we'd have to run the algorithm again and hope to not get 0 the second time around. If we read any of the other 3 values we'd be able to infer that $P = 4$. The easiest would be reading 64, which would allow us to directly compute 4, but 128 and 192 would work too with a bit of manipulation. Again, I'm glossing over some details. With $P = 4$ at hand, we can do some simple math using a classical computer to find the factors of 15, as follows:

$7^4 = m \times 15 + 1$
$7^4 - 1 = m \times 15$
$(7^2 - 1) \times (7^2 + 1) = m \times 15$ (This is why P has to be even!)
$48 \times 50 = m \times 15$

With the two factors 48 and 50, we can find the factors of 15 by using Euclid's algorithm, which is very efficient for calculating the greatest common divisors (GCD) between two numbers. GCD between 48 and 15 and 50 and 15 are 3 and 5, respectively. And with that, we finally arrived at our two factors that can break the public key 15: 3 and 5. Bada bing bada boom!

I know it was a lot of detail. I hope you didn't give up and are still reading this. The most important idea to retain from all of this is that the hard part, finding the period, is where the quantum magic happens. In classical digital computers, finding that period would take many tries. Much longer than a hacker may be willing to wait when trying to discover a private key. But a quantum computer can explore all these combinations at once using the quantum Fourier transform (QFT), like hearing a musical rhythm echo across all keys at once. Once we know that rhythm, the period, the rest is just normal math to extract the secret factors.

As we have seen, Shor's algorithm provides an exponential speedup for finding common factors and can be used to break RSA encryption. But don't worry, we are still a few years away from having working quantum computers with enough qubits to break any reasonably sized key. And even when we do, there will be alternative methods we'll be able to use to replace RSA, like quantum key distribution. Never doubt our ingenuity. We'll always find a way forward. With human agency, we can program our future. And very soon, we'll be able to do it in a quantum computer.

The physically-backed myth "our future"

By the time Albert Einstein died in 1955, Copenhagen was still the only accepted interpretation for quantum mechanics. Niels Bohr's **complementarity principle**, which states that we must be comfortable with classical physics governing particles and quantum physics governing the world of the small, was, and maybe still is, the most accepted worldview. Although Einstein wasn't comfortable with quantum non-locality and the measurement problem, no one doubts that math proposed by Heisenberg and others describes quantum behavior very accurately. And it has been the underpinnings of the huge success quantum mechanics has had in so many fields, from chemistry, to nuclear, to semiconductors. This success may be the reason physicists were encouraged to work on the practical problems, instead of following Einstein's footsteps and questioning its foundations.

In 1964, physicist John Stewart Bell, inspired by some of Einstein's experiments, proposed a famous theorem, **Bell's inequality**.[14] In physics, hidden variables are parameters that could describe the remote event, such as the remote collapse of the wave function in Einstein's box experiment. Bell's inequality is a mathematical constraint that holds true only if two assumptions are true. The first assumption is locality. No influence can travel

---

14. On the Einstein-Podolsky-Rosen Paradox., J. Bell, Physics Physique Fizika, 1(3), 195–200, 1964.

faster than the speed of light. One particle's measurement cannot instantly affect another distant particle. The second assumption is realism. Particles have definite properties, like spin or polarization, prior to measurement and hidden variables determine the outcomes of the measurement. A violation of Bell's inequality implies that at least one of these assumptions must be false. Either we must give up locality, or hidden variables, or both.

Bell's work brought new and renewed interest to the field of foundations of quantum physics, including the work that led to quantum computers. Subsequently, several experiments have disproved the hidden variables hypothesis and confirmed non-locality. The 2022 Nobel Prize in physics[15] was awarded to Alain Aspect, John F. Clauser, and Anton Zeilinger for their experimental work on entangled photons, establishing the violation of Bell's inequality and pioneering quantum information science. That work confirmed that quantum entanglement is real and not due to any classical mechanism or hidden signals limited by the speed of light. Although non-locality seems to be here to stay, today there are alternative interpretations to Copenhagen, like the many-worlds, which assumes that there is no collapse and outcomes happen in parallel branches. The quantum foundations field is flourishing again.

---

15. https://www.nobelprize.org/prizes/physics/2022/summary/ (Last accessed in May 2025).

If it was not for the questioning nature of people like Einstein and Bell, we might have never challenged the Copenhagen interpretation. Bell's work was key to bringing renewed attention to the field, eventually leading to the development of quantum computing. Challenging established worldviews is required if we want to see progress. This was a recurring theme throughout this book. Social networks, advertising, dataism, and all the systems that impact our lives can, and should, be challenged. This is why I stress that, like the Copenhagen interpretation, many of these systems and technologies are myths. Our lives are deeply influenced by technology. But we shouldn't be victims of technology. We can and should influence the direction it is moving.

The future of quantum computing is near. We are about to have the first large scale quantum processors. They will enable us to further advance research on basic science, including quantum physics itself. The initial quantum computers will be a great tool for developing more advanced quantum computer architectures and algorithms. As it turns out, quantum algorithms are great for modeling quantum physics. Richard Feynman described in his seminal paper[16] that classical computers are unsuitable to model quantum physics since the number of objects that need to be modeled scale exponentially

---

16. Simulating physics with computers, R. Feynman, International Journal of Theoretical Physics 21(6), 1981.

with the number of particles: n quantum particles require storing and processing 2^n numbers. As we've seen, quantum computers don't have this limitation. They are a natural fit for many problems that require simulation of nature, including quantum systems, molecules, and material science.[17]

With quantum, AI, cloud computing, and a lot of the technology we discussed throughout the book, we have tools to tackle our biggest societal problems, from health care, to housing, to climate change. The challenge is not technology. We need an engaged society, and institutions that enable us to get things done. Daron Acemoglu and Simon Johnson, argue in their book *Power and Progress*, that technological advancement, especially AI, is not inherently beneficial or harmful. Its effects on productivity, employment, and well-being depend on how it is deployed and who controls it. History shows that new technologies often concentrate power and wealth, unless society intervenes to steer their development toward shared prosperity. If innovation replaces more jobs than it creates, or primarily benefits a narrow elite, its net impact can be negative. Even as overall productivity increases. But this outcome is not inevitable. We have agency over how technology integrates into our lives. We can demand systems that enhance human capabilities, not

17. Disentangling Hype from Practicality: On Realistically Achieving Quantum Advantage, T. Hoefler, T. Häner, and M. Troyer, Communications of the ACM, May, 2023.

replace them. Acemoglu and Johnson propose that we focus on **machine usefulness (MU),** rather than intelligence. Rather than AGI, our quest should be for systems that promote dignity, not just efficiency. To make those choices, we must stay informed and actively engaged. It may make sense for AI to help us write code or drive a car, but perhaps not to raise our children or replace human connection. These are strategic choices, not technical ones. If we want a prosperous future, we must design it deliberately, starting now.

Conclusion

"I don't want to sit around and hope good things happen. I want to make them happen."

Drew Barrymore

As I was writing this book, my mom, alarmed by a Netflix documentary on consumerism,[1] told me I had to do something about it. The film focused on what Merriam-Webster defines as a preoccupation with or an inclination toward the excessive buying of consumer goods, driven by psychological manipulation and algorithmic targeting in online shopping and social media platforms. Consumerism fuels unsustainable spending, personal debt, and environmental harm, including resource depletion, waste generation and landfill overflow. I didn't know what to say. I felt both honored, that she somehow thought I had the power to change the consumer industry's influence on how people behave, and hopeless, knowing that any efforts on that front would likely be fruitless. All I could really do is what the documentary was already doing, give people the information, arming them to act. I also told her that I was already doing something by never "wanting anything as

---

1. https://en.wikipedia.org/wiki/Buy_Now!_The_Shopping_Conspiracy (Last accessed in March 2025).

birthday and Christmas gifts," which is something that makes her furious, and therefore, my comment was not very appreciated.

This book is my attempt to arm people with knowledge to reason about the systems that are so pervasive and so impactful to our lives. I leveraged the framework of myths proposed by Yuval Noah Harari to highlight that many of these modernities that seem scary to many, such as social media, self-driving cars, and AI, are just myths. We can, and should, understand them deep enough to have a productive conversation about their pros and cons. Knowledge is the first step towards action. And like my beloved Drew Barrymore said, action is better than sitting around and complaining about things.

Throughout the book I've described a series of technologies that I believe need our attention, as a society. Take self-driving cars as an example. My personal opinion is that a concentrated effort to make its adoption pervasive would generate so much societal benefit that it would be our modern day Manhattan project or space program. We'd see a reduced number of deaths and accidents, a positive impact on pollution and carbon emissions, less traffic, reduced transportation costs, and more beautiful cities, in which parking lots could be transformed into green communal areas. It would also produce infrastructure jobs and strengthen our investments in computational platforms, including cloud services and AI. Finally, it would also positively impact our housing crises, as with less traffic, the commute between geographically distant cities would improve.

I've also discussed technologies that I believe have a negative impact in society, such as some uses of social media. We must remember social media applications are a digital-only myth. We can, and should, think more thoroughly about the reasons we are allowing it to be so pervasive in our lives. I'm glad Jonathan Haidt and others are already leading this debate. In this book, we covered some of the technical reasons that make applications fragile. They don't even do what they were supposed to do properly.

Our world is becoming increasingly digital. However, most of our leaders, both in government and the private sector, do not understand technology at a sufficiently complex level. This disconnect is getting progressively worse. We are in the middle of the AI revolution and on the verge of the quantum computing revolution. These technologies are already drastically impacting how we work and how our society operates. We should have leaders that are worried about what we should be teaching to our children and how they will be collaborating in projects twenty years from now. There are incredible success stories when government leaders versed in technology invest in projects that promote social change.

In recent years, the Brazilian Central Bank, led by Roberto Campos Neto, implemented a country-wide real-time digital payments solution, called Pix. If you are familiar with Zelle, or Venmo, Pix is similar. The main difference is that the government provides it at no cost. Since its inception, it enabled millions of people to be enfranchised into the Brazilian financial system. Currently, 85% of the population has access to financial services. The Gates Foundation also provides success stories on how technology can be used to solve the world's greatest problems, such as the eradication of malaria and combating the spread of HIV. It starts with understanding these problems and understanding technology to solve them.

One of the most relevant books I read recently is *Algorithmic Institutionalism*.[2] It expands our current view of institutions to also include algorithms, given the impact they have dictating behaviour in our lives. Computer systems, as we've discussed, influence how to work, shop, and read the news. One of the many examples discussed in *Algorithmic Institutionalism* is in the healthcare industry, specifically, choosing recipients for kidney transplants. There are several rules encoded in the system that may have a large impact on what individuals on the transplant list receive the donated organs. As the authors point out, algorithms play a central role in our society and should be treated as such. We need better cooperation between society, government, and the private sector, jointly thinking about projects and regulations that will enable us to design the future we want and deserve.

---

2. *Algorithmic Institutionalism: The Changing Rules of Social and Political Life*, R. Mendonca, F. Filgueiras, and V. Almeida, Oxford University Press, 2023.

I wrote my previous book, *A Platform Mindset,* as I identified a foundational gap in technology leaders. Most of them, myself included, started as developers and progressed through the ranks. As stated by our favorite business theory, the Peter Principle, we grew to our maximum level of incompetence. *A Platform Mindset* was my attempt to disseminate the lessons I've learned during my career to new tech leaders. This book was a bit more ambitious. I wanted to democratize the understanding of technology, so that we can have more informed conversations about the societal changes we want to promote in this digital age. With tech billionaires more interested in colonizing Mars and achieving AGI than in solving our current problems, we need the rest of society and government to engage in understanding, regulating, and guiding the development of algorithms and computer systems.

I think it is much more important to solve our current problems, access to energy, climate change, drug discovery, transportation, than to worry about life on Earth in one thousand years or if we'll achieve AGI or not. With or without AGI, we already have the tech we need to address these important societal issues. But we need engagement from society. In their insightful book,[3] Ezra Klein and Derek Thompson discuss the politics of abundance. How to continue growing the economy by investing in and delivering projects that will address our

---

3. *Abundance,* Ezra Klein and Derek Thompson, Avid Reader Press / Simon & Schuster, 2025.

basic needs? It is better to fund projects that will cause breakthrough advances in medicine and cost of living than to try to fix the budget for Medicare and social security. However, if we want to make progress on these projects, the future will not be like the present. The better we understand the technologies shaping our lives today, the more we will be able to guide our future. We need human agency to drive the need for machine usefulness (MU), rather than intelligence or automation. We need to understand the algorithms that are so present in our lives, so we don't get scared by them. Instead, we should understand the benefits we are getting by adopting them and we have the tools to make the right decisions.

One important discussion in this book is the tradeoff between efficiency and humanity, as in the case of South Korea, where we are now seeing the downsides of a hyper efficient society, such as mental health and low birth rates. If I had to summarize this tradeoff in a single sentence, I'd say that we need efficient systems to tackle the problems that will have a positive impact on society. However, we need to be very diligent about when to use computer systems, and we need to be careful about technology adoption. If we need to build a sorting system let's use quicksort or bubble sort, instead of bogosort. But, before doing that, let's try to evaluate if the system will enrich our human experience. Efficiency for the sake of efficiency may lead to unwanted consequences.

I hope that this book will also inspire younger generations to learn more and take an active role in shaping our future. We need well-rounded leaders that understand technology and how it can be used to impact society in a holistic way. Nonprofits like Code.org play a central role in disseminating access to computer science education. And we should do more, given the omnipresence of computer systems in our lives. With that in mind, all proceeds from this book will be donated through Microsoft Philanthropies to organizations invested in the democratization of computer science education.

My teenage daughters are both worried about AI. They keep asking me if it is possible that robots will kill humanity. On a more serious note, they also ask me what people will do in the future when AI can perform all tasks better than humans. We are still in the early stages of the AI revolution. If we invest in understanding technology better, and if we require that our leaders do the same, we can influence our future. It is in our hands. But we must all do our part. Let's act together to build a better, more efficient society where we are productive and happy. I'd love to hear your thoughts about the book and continue the dialogue at https://digitalagencybook.org.

Glossary

- **10x developer:** a software developer who is significantly more productive and effective than an average developer. These individuals exist, but the company environment, team dynamics, and company culture also play vital roles. A 10x engineer's productivity is not solely about individual effort but also about leveraging the organizational environment and platforms.
- **10-90 decisions:** decisions where one option is significantly worse than the other.
- **40-60 decisions:** decisions where one option is slightly better than another, but both are reasonable.
- **Abstraction:** representations that allow developers to model complex real-world or digital concepts using the simple computer language of zeros and ones. They allow developers to think about things like individual numbers, lists, matrices, websites, songs, and videos, rather than sequences of bits. In the same way culture evolves faster than biology, abstractions evolve much faster than computer hardware.
- **Algorithm:** a step-by-step procedure or set of rules to solve a problem or complete a task.
- **Algorithmic institutionalism:** the integration of algorithms and computer systems into the core processes and structures of our society. Algorithms behave like

institutions, including organizations, governments, and even social frameworks.

- **Anchor text:** the text in and around a hyperlink. Anchor text plays a significant role in web search, as it provides context to the linked page, helping search engines understand what the linked content is about. It influences search rankings.

- **Antifragile:** a system that benefits from stressors, shocks, volatility, noise, mistakes, faults, attacks, or failures. As discussed, modern AI training exemplifies antifragility. With more data, the system improves.

- **Artificial general intelligence (AGI):** a hypothetical system that is generic, not requiring external specialized tools, and can perform any intellectual task that a human being can.

- **Attention:** the mechanism in transformer models that allows them to focus on salient parts of the text. This is crucial for understanding the relevant context in a prompt.

- **AutoCorrect:** a software function that automatically corrects spelling mistakes as text is typed.

- **AutoComplete:** a software feature that predicts and suggests words or phrases as the user types.

- **Autoscaling:** the process of automatically adjusting the number of virtual machines (VMs) or other computing resources allocated to a service based on its current demand or load. When traffic increases, autoscaling adds more VMs to handle the load, and when traffic decreases, it removes them to reduce costs. This ensures that the service has the necessary resources to operate efficiently during peak times while minimizing waste during off-peak hours, allowing the system to adapt to changing user demands dynamically and maintaining high performance.

- **Backend:** a **cloud service** that performs a specific function. Returning a cohesive search results page, for instance, requires calling several backends, including web search, ads, maps, and Wikipedia.

- **Barrier of entry:** the significant obstacles that make it challenging for individuals to enter a market, such as becoming a successful published author or a NFL quarterback.

- **Big-O:** a notation used in computer science to describe the performance or complexity of an algorithm.

- **Bits:** the term stands for **b**inary dig**its**, the smallest unit of information in a digital system. Each bit is either 0 or 1, represented physically in modern CPUs and GPUs using transistors as tiny electronic switches that can be on or off. 0 can be represented by 0 volts while 1 is represented by supply voltage. Logic gates process these voltage levels to implement all possible functions computers perform.

- **Black swan event:** an unpredictable event that is beyond what is normally expected and has potentially severe consequences. In the context of computer systems, a black swan event refers to a sudden and extreme increase in demand or usage that a system is not designed to handle, leading to potential failures or outages. Unlike physical systems that are constrained by well-understood physical laws, computer systems are more vulnerable to these unpredictable events due to their potential for exponential scaling in use.

- **Bogosort:** a highly inefficient sorting algorithm that randomly shuffles a list until it is sorted, used as an example of inefficiency throughout the book.
- **Bubble sort:** a simple sorting algorithm that repeatedly steps through the list, compares adjacent elements, and swaps them if they are in the wrong order.
- **Byte:** a unit of digital information that most commonly consists of eight bits.
- **Central processing unit (CPU):** the primary component of a computer that executes instructions of a program. I differentiated CPUs from GPUs, particularly in their role in handling the parallel processing demands of modern AI models, such as those in the transformer architecture. CPUs aren't as efficient as GPUs for the massively parallel computations that are required for modern AI.
- **Chain-of-thought (CoT) reasoning:** a technique used with AI models where the model is asked to produce an explanation before producing the answer, breaking down the problem into smaller steps. This is likened to activating "System 2" thinking in humans.
- **Chaos engineering:** intentionally introducing failures into a system to test its resilience and identify weaknesses. By injecting faults, such as server crashes or

network disruptions, chaos engineering reveals how a system responds to unexpected events.

- **Cloud computing:** using remote data centers to run virtual machines (VMs) instead of local physical servers. These VMs are accessible over the internet, and users can interact with them in the same way they can with servers located under their desks. Cloud computing allows for scalability, resource sharing, and access to computing power without the direct management of hardware.

- **Cloud service:** services provided over the internet, allowing users to access computing resources and applications without direct management of the underlying infrastructure.

- **Compression:** reduce data size to save storage space or transmission time, as mentioned with respect to Huffman code.

- **Context:** the surrounding information that helps provide meaning to a word, phrase, or situation. In AI, it refers to the amount of information a model can consider when processing input.

- **Context size:** in AI models, the number of tokens the model can consider when processing input. GPT-3, for instance, has a context size of 2,048 tokens.

- **Copenhagen interpretation:** an interpretation of quantum mechanics that sees measurement as collapsing the wave function, championed by Niels Bohr, as discussed in the context of Einstein's concerns about non-locality.

- **Cores:** a processing unit within a central processing unit (CPU). Multi-core processors have multiple cores, allowing them to execute multiple instructions simultaneously, thereby increasing processing power and speed. GPUs also provide a larger number of cores that can be used to run computations in parallel.
- **De-embedding matrix:** a matrix that maps embedding dimensions back to the token space, used to determine probabilities for the next token.
- **Deterioration:** a decline in quality or efficiency of organizations, often combated by investing in technologies that produce efficiencies, such as spell checkers and GenAI.
- **Deterministic:** an algorithm that produces the same output whenever called with the same input.
- **Dewey decimal system:** a classification system traditionally used to categorize books in libraries that can be used to represent index positions for a search engine. Rather than simply storing page numbers or raw positions in a document, Dewey numbers provide a more structured and hierarchical representation of location. For example, a Dewey number could represent a chapter, section, paragraph, and sentence. This encoding of more information into the position data allows for more precise and nuanced search queries, such as finding matches within the same chapter, section, paragraph, or even sentence. While the Dewey system was originally devised to organize libraries, in this context it is used as a way to represent digital index

locations, enabling more complex search functionalities at the cost of more complex index representation and compression.

- **Digital agents:** software systems that automate tasks using GenAI models, like a system that continuously scans the new sites and keeps you updated, sending you summary emails as soon as news that match your interest are published.

- **Digital transformation:** the integration of digital technology into all areas of a business, fundamentally changing how it operates and delivers value to customers. The e-book and VMs are examples of digital transformations for books and physical servers.

- **Digital-only myth:** myths sustained purely by shared belief in the digital realm, such as online advertising and social networks.

- **Dimensionality reduction:** the process of reducing the number of dimensions of data, mapping from a high-dimensional space to a lower-dimensional one. Early AI relied heavily on dimensionality reduction, limiting the amount of information used but making it easier to explain how the system worked.

- **Distributed systems:** systems composed of multiple independent computers that communicate through a network to achieve a common goal. Rather than relying on a single machine, distributed systems leverage the combined resources of many machines. This approach enables greater scalability, fault tolerance, and performance. We discuss distributed systems as key for large-scale search engines and cloud computing, which require massive computing power and the ability to handle enormous amounts of data and traffic.

- **Document sharding:** splitting of a large collection of documents, such as web pages for a search engine, into smaller, independent subsets called "shards." This division allows for the storage and processing of these documents across multiple servers, enabling parallel processing and reducing the load on any single machine. This is essential for handling the vast scale of data in web search and provides for the ability to do many computations at the same time. Document sharding allows for faster data retrieval, efficient indexing, and improved scalability of the search engine.

- **Dot product:** a mathematical operation between two vectors that results in a single scalar value. Specifically, it's mentioned in the context of how GPT models perform matrix multiplications, where these operations increase or decrease the "weights" in embedding vectors. These weights represent the significance or contribution of different dimensions to the meaning of each token. The dot product, in this sense, is a key operation that alters the vectors representing tokens during self-attention and feedforward stages.

- **Drew Barrymore:** cherished actress, who is hard to search for, but regardless of the character's memory loss, always manages to have a happy ending, especially if there are tapes involved.

- **Embeddings:** vector representations of words or tokens in a high-dimensional space, capturing both semantics and syntax. GPT-3, for example, has 12,288 dimensions per token embedding.

- **Efficiency:** a measure of how well a system is performing. We introduced many definitions of efficiency throughout the book, such as order, or Big-O, PUE, and utilization. We discussed efficiency in the context of algorithms, systems, and personal usage efficiency, or productivity.

- **Endogeneity:** in an economic context, this refers to situations where factors within a system are influenced by other variables within the same system. For instance, the higher SAT scores may not come for the students that studied the most, simply because they may not have needed to study as hard.

- **Energy proportionality:** systems where energy consumption scales proportionally to the amount of work done. Ideally, a system should consume minimal energy when idle and only increase consumption as the workload increases.

- **ENIAC:** Electronic Numerical Integrator and Computer (ENIAC), which was the first general-purpose electronic digital computer.

- **Entanglement:** in quantum mechanics, when two particles are linked together so that measuring one instantly affects the other. Called "spooky action at a distance" by Einstein.

- **ESP game:** an online game designed to produce labels for images, used to train AI systems.
- **Factorization:** the process of breaking down a composite number into its prime factors, numbers that can only be divided by 1 and themselves. RSA security depends on the computational difficulty of factoring large numbers. Shor's algorithm demonstrates that quantum computers can solve this problem efficiently, meaning that quantum computing could break RSA.
- **Fault domain:** in cloud computing, a fault domain refers to a section of an infrastructure that can fail independently of others. A common strategy to achieve high reliability is to run multiple replicas of a service across different fault domains, so that if one domain fails, the service can continue to operate using the other replicas.
- **Features:** in AI, distinctive characteristics or attributes of data that are used for prediction or classification. In old school AI, feature selection was an "art" requiring human expertise.
- **Fragile:** a system that is harmed by stressors, shocks, volatility, noise, mistakes, faults, attacks, or failures.
- **Generative AI (GenAI):** artificial intelligence that is used to generate new content, such as text, images, or code.
- **Graphics processing unit (GPU):** initially designed for accelerating the creation of images and videos, GPUs have evolved to be highly efficient at handling parallel computations. They consist of thousands of smaller, more specialized cores that can work on multiple tasks simultaneously, making them well-suited for tasks like

training and running large AI models. GPUs provide a larger number of specialized processors that can be used to run transformer computations in parallel, allowing AI applications to answer complex prompts quickly. This parallel processing capability significantly speeds up the matrix multiplications and other operations required by these models.

- **Grover's algorithm:** a quantum algorithm for searching an unsorted database with quadratic speedup over classical algorithms, reducing the steps from N to square root of N.

- **Hallucination:** GenAI has no idea what it is writing about. Or more precisely, when an AI model generates false or nonsensical information. There is nothing in training or inferencing that protects AI systems from producing a completely random response.

- **Herding:** see information cascades.

- **h-index:** a metric used to quantify a researcher's productivity and impact. A scientist has an index h if h of their N papers have at least h citations each, and the other $(N-h)$ papers have no more than h citations each. The h-index is used as a measure of academic impact and influence.

- **Huffman code:** a data compression algorithm that assigns variable-length codes to input characters based on their frequency of occurrence.

- **Human agency:** the capacity of individuals to act independently and make their own free choices. It is about understanding, evaluating, and using technology in a way that empowers humans, or not using it if it is harmful. It emphasizes that humans are autonomous beings who act by themselves and independently, even within a world of digital agents.

- **Human computers:** before electronic computers were widespread, human computers were individuals who performed complex mathematical calculations by hand. The work of human computers was crucial for scientific and technological advancement in the past, and their role highlights the shift from manual computation to automated processing. They are the exact role that was replaced by the ENIAC and following machines.

- **Hyperplane:** a generalization of a plane to higher dimensions, used to separate data points in a high-dimensional space.

- **Information cascades:** the phenomenon where people make decisions based on the actions of others, rather than their own private information.

- **Inferencing:** using a pre-computed AI model to answer queries or generate outputs. For text, inferencing means computing the next words.

- **Innovator's dilemma:** the challenge faced by established organizations when disruptive technologies or new business models emerge. These innovations may initially seem less profitable or less relevant to existing customers, leading established companies to ignore them in favor of sustaining existing offerings.

- **Interactive deployment:** a method used by companies like OpenAI for releasing new AI systems that minimizes negative impact by gathering early feedback, which is then used to refine the models.

- **Intersection algorithm:** used to find the common elements between two or more sets of data, such as lists from an inverted index used in web search.
- **Inverted index:** a mapping from words or tokens to their positions, often used in search engines.
- **Knapsack:** the problem of selecting the most valuable items (tasks, processes, or resources) to include within a limited capacity (time, budget, or computational power).
- **Kicking a computer:** not recommended. It is very unlikely it will fix your code.
- **Large language model (LLM):** a type of AI model trained on vast amounts of text data to understand and generate human-like language. LLMs have many parameters and a high-dimensional embedding space.
- **Lean manufacturing:** a production philosophy focused on maximizing customer value while minimizing waste. It originated from the Toyota Production System (TPS) and has since been widely adopted across industries far beyond automotive.
- **Linear regression:** a statistical method used to model the relationship between a dependent variable and one or more independent variables by fitting a linear equation to observed data. Used to predict service growth, as an example.
- **Logic gates:** logic gates are fundamental building blocks of digital circuits that perform basic logical

operations. Logic gates such as AND, OR, and NOT work with binary inputs (0s and 1s) to produce binary outputs, thereby enabling complex digital functions to be built up from simple elements.

- **Machine usefulness (MU):** a concept proposed by Daron Acemoglu and Simon Johnson in their book, *Power and Progress,* based on the idea that we should be more concerned with machine usefulness than intelligence. Useful algorithms and systems are intended to enhance human capabilities and generate new, productive tasks rather than primarily eliminating human work.

- **Myth:** in this book, I use the definition of myth from Yuval Noah Harari's *Sapiens.* Complex concepts have no real representation in the real world, such as cities, companies, and systems like social networks.

- **Normal distribution:** a bell-shaped curve where most things are clustered around a middle number, with fewer occurrences on either side. It is characteristic of distributions found in the natural world, such as the height and weight of humans, and amino acid frequencies in proteins.

- **NP-hard:** a class of problems that are at least as hard as the hardest problems in NP (nondeterministic polynomial time). Often contrasted with problems that have efficient solutions. We discussed **knapsack** as an example of an NP-hard problem.

- **Online advertising:** using the internet to deliver promotional marketing messages to consumers.

- **Oversubscription:** in cloud computing, oversubscription is the practice of allocating more cores or memory resources to VMs than are physically available on the underlying hardware. Cloud providers often oversubscribe resources because not all VMs require their allocated resources at the same time. By oversubscribing, they can increase the utilization of their physical servers and reduce costs.

- **PageRank:** algorithm used by web search systems to explicitly rank pages based on their link structure. It operates by analyzing the citation-like structure of links between web pages, assigning higher ranks to pages that are linked to by many other pages or pages that are themselves highly ranked. This differs significantly from how Large Language Models (LLMs) operate, as LLMs do not have built-in schemas for source credibility.

- **Parameters:** the internal settings of an AI model, such as weights in matrices, that are learned during training. GPT-3 has 175 billion parameters.

- **Peter Principle:** the observation that people in a hierarchy tend to rise to "their level of incompetence," often used to illustrate inefficiencies in organizations.

- **Personal usage efficiency (new PUE):** A new spin on the traditional power usage efficiency applied to productivity. Defined as total work divided by total value-producing work (VPW). Lower PUE values represent more efficient organizations.

- **Physically-baked myth:** myths anchored in physical reality, such as companies and cities.
- **Placement constraints:** rules or policies that dictate where virtual machines (VMs) or other computing resources should be allocated within a data center. These constraints are often used to ensure reliability and performance, such as placing VMs across different **fault domains** or on specific hardware configurations.
- **Platform investment:** investing in technologies that produce efficiencies, such as spell checkers and GenAI, and then reinvesting the gains to fund technologies that will combat disruption, such as new payment methods and online shopping tools. This creates a virtuous cycle for organizations.
- **Power law distribution:** distributions characterized by a "long tail" instead of a bell curve. These distributions are common in human-created concepts like letters in English texts, city populations, and wealth, and are generally harder to predict and rationalize than normal distributions. Computer systems and financial markets are more prone to black swan events due to the presence of power law distributions.
- **Power usage efficiency (PUE):** a metric used to measure the energy efficiency of data centers. It is calculated as the total facility energy divided by IT equipment energy. A lower PUE indicates better efficiency, meaning more of the total energy is used for computing rather than overhead like cooling.

- **Principle of locality:** the concept that all actions and their side effects are local, which is violated by quantum entanglement.
- **Quantum gates:** quantum gates are the building blocks of quantum circuits. Analogous to logic gates in classical computing, quantum gates operate on qubits (quantum bits) to perform quantum operations. These gates can manipulate the superposition and entanglement of qubits, enabling quantum algorithms to solve certain problems more efficiently than classical algorithms.
- **Qubits:** the basic unit of information in a quantum computer, which can be in a superposition of 0 and 1.
- **Reinforcement learning from human feedback (RLHF):** a technique where human feedback (e.g., ratings) is used to refine AI models.
- **Return on Investment (ROI):** A measure used to evaluate the efficiency of an investment. As described in the context of chaos engineering, ROI is calculated by subtracting the investment in chaos engineering, the chaos budget, and the new incident cost from the original cost of incidents before adopting chaos engineering, then dividing that return by the total investment.
- **Rightsizing:** adjusting the size and configuration of virtual machines (VMs) to match the actual needs of the services running on them. This includes selecting the appropriate number of CPU cores, memory, and storage. Rightsizing ensures that VMs are not over-provisioned, wasting resources, or under-provisioned, leading to performance issues.

- **Robotaxi:** a self-driving taxi service. These vehicles operate autonomously, without a human driver, and can be summoned via an app or service. They represent an advancement in transportation, promising increased efficiency, reduced accidents, and potential changes in urban planning and personal car ownership.
- **RSA (Rivest-Shamir-Adleman):** a public-key cryptography system widely used for secure data transmission. Its security relies on the difficulty of factoring large numbers into their prime factors. Shor's algorithm can efficiently perform this factorization, which could break RSA encryption. This makes RSA a prime example of encryption methods vulnerable to quantum computing advances.
- **Salience:** information stored in search system lists that indicates the prominence of a term's occurrence within a document. This can include whether the term is in bold, italics, or in a header, and is used to compute a similarity score for search results.
- **Self-attention:** a mechanism in transformer models that allows the model to focus on different parts of the input sequence when processing each word.
- **Self-driving cars:** Self-driving cars, or autonomous vehicles, are vehicles capable of navigating and operating without human intervention. We used it to exemplify both an antifragile system as it learns with every mile driven and a physical system becoming more digital over time.

- **Server:** a powerful computer designed to provide data or services to other computers over a network. It can be physical hardware or a virtual machine (VM) running in a remote data center. Historically, servers might have been placed under a user's desk, but now they are typically remote.
- **Shor's algorithm:** a quantum algorithm for integer factorization, which has exponential speedup over the best-known classical algorithms and can break RSA encryption.
- **Social networks:** online platforms where users create profiles, connect with others, and share and interact with content. They are susceptible to information cascades.
- **Sorting:** arranging items in a specific order, often used as an example of different algorithms' efficiency, such as **bubble sort** or **bogosort**.
- **Statistical machine translation (SMT):** machine translation systems that use statistical models based on large amounts of bilingual text to learn translations.
- **Stopwords:** very popular terms like "the" or "a." These terms are generally irrelevant for search systems, and are many times disregarded completely, as their presence won't impact the results.
- **Superagency:** a concept proposed by LinkedIn cofounder Reid Hoffman and his coauthor, Greg Beato, in their book of the same name, where AI and digital agents improve human agency to a new level, allowing for better and more efficient decision-making.

- **Superposition:** in quantum mechanics, the principle that a quantum system can exist in multiple states simultaneously until measured.
- **System 1:** in cognitive theory, the fast, automatic, intuitive, and unconscious mode of thinking.
- **System 2:** in cognitive theory, the slow, analytical, and effortful mode of thinking. In this book, I associated chain-of-thought reasoning with activating system 2 in GenAI systems.
- **System prompt:** directions or traits provided to an AI model along with the user's prompt to guide the model's output.
- **Temperature:** a setting in some GenAI models that influences the randomness and creativity of the output. A higher temperature allows the inferencing algorithm to select tokens with lower probabilities, leading to more creative but potentially less accurate or "hallucinated" outputs.
- **Term sharding:** dividing the index of terms used in a search engine. Instead of a single index containing all terms, the index is split based on the terms themselves. For example, terms starting with "A" might be on one server, and terms starting with "B" on another, and so on. This allows the system to handle a very large number of unique terms.
- **Timesharing:** the technique of dividing processing time among multiple programs or users. This creates the illusion of multiple programs running simultaneously, even though only one program is actively being

processed at any given instant. It enhances efficiency and overall system utilization, as the processor is not left idle while waiting for a single program to complete.

- **Tokens:** basic units of text processed by AI models, which can be words, subwords, or combinations of words.
- **Training:** The process of teaching an AI model by feeding it large amounts of data.
- **Transformer block:** A component of transformer models consisting of a self-attention step followed by a feedforward step.
- **Transformers:** An AI technique that allows for parallel processing of input data and focuses on salient parts of the text through attention mechanisms.
- **Uniform superposition:** A state in quantum computing where all possible states have equal probability.
- **Value-producing work (VPW):** a concept from lean manufacturing that indicates work that directly contributes to the intended outcome or purpose of an organization.
- **Vibe coding:** term used for building software with the help of GenAI tools. It greatly improves productivity in many scenarios but often requires human assistance. GenAI helps experienced developers more than inexperienced ones.
- **Virtual machine (VM):** a software emulation of a physical computer. It provides a complete computing environment, including an operating system and applications, that runs on top of a physical server. VMs offer flexibility,

scalability, and efficiency, as they can be easily created, resized, or destroyed on demand, which contrasts greatly with the physical world and purchasing new servers.

- **Weak ties:** in social networks, weak ties refer to connections between individuals who are acquaintances or peripheral contacts, rather than close friends or family. While strong ties provide emotional support and reinforce existing beliefs, weak ties often act as bridges to diverse information, opportunities, and perspectives outside one's immediate social circle. These ties are critical for the flow of novel ideas and information, as they connect disparate clusters of people. Weak ties play a valuable role in job searches, finding new opportunities, and accessing diverse information, leading to greater efficiency in the spread of information in a larger network.

- **Web search:** using algorithms to rank and retrieve information from the vast amount of content available on the internet. Unlike large language models (LLMs), traditional web search systems often explicitly rank pages based on link structure and credibility, using algorithms such as PageRank.

- **Y-career path:** development model in organizations where employees can choose to advance either on a managerial track or an individual contributor track. This allows skilled professionals to stay focused on their technical expertise without necessarily needing to move into management roles to progress in their careers and receive promotions, as is the case with the Peter Principle.

Acknowledgments

After I wrote *A Platform Mindset*, several readers thanked me, saying how much they appreciated that I was able to translate difficult technical concepts into a simple language. One comment that stuck with me was, "you were able to democratize and humanize technology." Those comments encouraged me to write this book. I'm very grateful for the positive feedback and for the encouragement to continue this journey.

This book wouldn't be possible without the help of many people. My friends and colleagues Kiana Baharloo, Priscila Bortoloto, Eli Cortez, Sekhar Pasupuleti, Lucio Tinoco, Zuzu Tang, Matthias Troyer, and Brijesh Warrier, did a very thorough and insightful review of early drafts. Their revisions were fundamental to improving the book. Kiana did a very detailed review of the manuscript and proposed several improvements that greatly improved the quality and consistency of the text. Prisicila found several typos during her detailed reading, even when I thought the book was polished enough. Eli pointed me to several resources that helped me describe how LLMs work and provided many insightful suggestions in his detailed reviews. Sekhar proofread an initial draft, providing many comments that greatly improved the book. Lucio took the time to propose several improvements to the text, such as using the insightful piano key analogy to describe Shor's algorithm, amongst many others. Matthias did a very thorough review of the quantum computing chapter and provided many insightful comments that greatly improved it. Zuzu proofread the book in detail and proposed beautiful modifications to some concepts and algorithm descriptions. Brijesh was very enthusiastic about the book, provided very insightful comments, and contributed ideas to the book title.

I'm also extremely grateful to Michael Schwarz, for the thoughtful *Foreword*. Michael took considerable time to read the book in detail and we discussed several topics at length, which greatly improved the final version. Thank you also Virgilio Almeida, Girish Bablani, Luiz Ceze, Don Cowan, Hardi Partovi, Brad Porter, Kira Radinsky, and Matthias Troyer for the comments, encouragement, and for the endorsements.

Thank you to Greg Shaw, Steve Clayton, Barbara Byrne, Megumi Voight, Myles Thompson, and the team from 8080 Books for the support and hard work on this project. Greg in particular, as he was the first one to read an initial version of the first few chapters. His early feedback and guidance completely changed the framing for the book. Mark Fortier, Victoria Comella and the entire team from Fortier PR for believing in this project and helping position the book.

Grant Sanderson, who I don't personally know, has an amazing YouTube channel that uses animations to describe scientific concepts, 3Blue1Brown.[1] I relied heavily on his videos for some of the descriptions about large language models (LLMs), especially embeddings, and quantum computing. I strongly recommend the videos, as the animations can help clarify some of the key concepts.

My family was extraordinarily understanding and supportive throughout the writing of this book, even when it meant long hours away from them and missed moments together. I still remember the look on my daughters' faces when they saw me typing the initial words on a document named *A book about efficiency* and asked "Are you going to write another book?" Their patience gave me the space to think and create, and their encouragement sustained me during the inevitable periods of doubt and fatigue. I am deeply grateful for their unwavering belief in me, for reminding me of what truly matters, and for grounding me when the work seemed all-consuming. This book could not have been written without their love, generosity, and constant support.

1. https://www.3blue1brown.com/ (Last accessed May 2025).

About the author

Marcus Fontoura is currently in his second tenure as Technical Fellow and Corporate Vice President at Microsoft, where he works as CTO for Azure Core. He is also the author of *A Platform Mindset: My lessons from developer to CTO*, published in 2025 by 8080 Books.

Most recently, he was the CTO at Stone.co (2022-2025), where he led the engineering organization, focused on building highly efficient financial platforms and an amazing engineering culture. He continues to serve as an advisor to the company.

Previously, in his first tenure as Technical Fellow and Corporate Vice President at Microsoft (2013-2022), Marcus worked as the chief architect for Azure Compute and led the Azure efficiency team. Marcus also had posts at Google, Yahoo, and IBM Research. He was a post-doctoral researcher at Princeton University and has received his Ph.D. in Computer Science by the Pontifical Catholic University of Rio de Janeiro, in Brazil (PUC-Rio).

Other titles from 8080 Books

No Prize for Pessimism by Sam Schillace
WorkLab by Colette Stallbaumer
A Platform Mindset by Marcus Fontoura

8080
BOOKS

Check for new titles on our website:

https://unlocked.microsoft.com/8080-books/